MW00974430

Enjoy these other books in the series:

We Were Crewdogs I – The B-52 Collection
We Were Crewdogs II– More B-52 Crewdog Tales
We Were Crewdogs III – Peace Is Our Profession
We Were Crewdogs IV – We Had To Be Tough
We Were Crewdogs V – We Flew the Heavies
We Were Crewdogs VI – Freedom Is Not Free
We Were Crewdogs – The Vietnam Collection

AND

Linebacker II – A View from the Rock (Revived)

Moments of Start Terror!

Printed versions available by mail from
www.wewerecrewdogs.com

Tommy Towery
5709 Pecan Trace
Memphis, TN 38135
tommytowery@gmail.com

Cover artwork by Rock Roszak, Col USAF (Ret.)

eBook versions available from
Amazon.com

ISBN: 978-0-9854108-4-1

The B-52 Factor

We Were Crewdogs VII

Edited by
Tommy Towery

Table of Contents

Stories by Authors

Randy Wooten

Jim Wuensch

John York

Factor [fak-ter] - *noun* – something that helps produce, influence, or contributes to a result, process, or accomplishment.

Me at U-T

Introduction

Tommy Towery

In the beginning, God created the heavens and the earth - and CINCSAC. And CINCSAC said, “Thanks. You’re dismissed. I’ll take it from here.”

As powerful as he was, eventually the Commander in Chief, Strategic Air Command (CICSAC) went away, as did the Strategic Air Command (SAC) itself. But not so the mighty B-52 Stratofortress. The giant eight-engine bomber, first flown in 1955 and revered backbone of our national defense during the Cold War, still remains a viable weapon system in today’s United States Air Force’s inventory. It shall be for years to come. But the aircraft itself is useless without the men who built, guarded, maintained, and flew the plane affectionately known as the Big Ugly Fat Fellow (BUFF) by many. Okay, maybe the final “F” really stands for something else for all those who really know her, but you get the point.

This book is the seventh volume of stories primarily about the aircrews who flew the plane but is dedicated to all those who ever came in contact with one. Despite a half-dozen efforts to recruit someone to take lead on a book about the maintainers of the aircraft, no one has been willing to come forward and take on the task, so I will continue to publish stories about the aircrews. I was never a maintenance officer; I was a flyer. I only really know and can vouch for things affecting the lives and events of the aircrew (Crewdogs) who flew her. There are a few story exceptions in this volume, but those stories included help add to the understanding of what we did, and deserve to be heard.

Sue, my wife and an avid reader, came up with the name for this volume in the series – *Volume VII -The B-52 Factor*. The definition of factor is "something that helps produce, influence, or contributes to a result, process, or accomplishment." And that is the foundation of all the stories included. This is not a book about the B-52 aircraft itself or its capabilities, but about the factor it played in the lives of the crews who manned it.

You may remember George C. Scott's portrayal of Gen George S. Patton's in the movie *Patton*. He addressed the troops saying, "Thirty years from now when you're sitting around your fireside with your grandson on your knee, and he asks you: 'What did you do in the great World War II?'…." My father was in World War II and never sat me on his knee and told me about his combat or even his training experiences. It took years for me to learn he was a member of the 29th Infantry Division and landed with them at Omaha Beach on D-Day, June 6, 1944. Oh, I knew he was in the war and was a disabled veteran, because he was missing his lower left leg, the aftermath of stepping on a land mine as he charged on the beach. He had a Purple Heart Medal he kept in a dresser drawer to show for it. That was about all I knew as a child. It was only through reading his newspaper interview for the 40th anniversary of the landing that I found out he lay on the beach for 18 hours before he was medically evacuated to a hospital ship. The story appeared while I was stationed in England near the end of my Air Force career and encouraged me to plan to sit down and make him talk to me about the events leading up to and following his war time experiences. He died before I got home to do so. By that time he had his other leg amputated, the result of the same battle injury he suffered in 1944.

That was the Genesis for my decision to try to document some of the things about my Air Force life and the lives of my fellow B-52 aircrew members before it was too late. If I could not understand something as fundamental as what an Army infantryman did, how could I expect any of my family to ever comprehend the complicated life of a B-52 Electronic Warfare Officer? I wanted to preserve a few of the things about those days so someday my family would know about what I did. I want my daughter to understand why I was gone so often when she was growing up. I didn't plan for it to be just a book about me, but I wanted to be able to give others a chance to share their stories as well and give the subject a little more in-depth analysis.

After *We Were Crewdogs I – The B-52 Collection* was printed, others contacted me saying they had a story they wanted to preserve and would I consider doing another book. For many, the memories documented in the *We Were Crewdogs* books are the only written record of their service time. Several former story contributors have flown west now and it may be the only printed story their families will ever be able to pass down to the next generation. This volume will add to the total of over 300 stories in seven books which tell the tales. Some are boring, some are exciting. All are true.

In the beginning it was my goal to allow each contributor to submit his story in his own words, the way he would tell his tale in person. I received some criticism for some of those stories because of the grammar and style used. With each additional book I have tried harder to preserve the style the best I can, but still polish the stories which could be edited without taking away from the real story or story teller.

The stories are what they are. As stated before, some are boring, but some of the things we did were boring. It was still a part of our crew lives. These are real stories by real Crewdogs, told as they want to tell them. The books will never be on a best seller's list, nor will they win a literary prize, but they do paint the best possible and truest picture of the daily lives of the men who flew the B-52. These are the stories the authors want their friends and family to remember.

I expect I speak for all the story tellers from all the volumes when I say I hope you enjoy them, and hope they do a good job of telling it like it was – on all occasions. I have tried to present them in the best possible way, but I still live in fear of the old adage, "To Err is Human, To Forgive is not SAC Policy."

Chapter One – Military Career

M**ilitary Career** [mil-i-ter-ee] [kuh-reer] *–noun* - An occupation or profession, especially one requiring special training, for, or pertaining to war - followed as one's lifework.

A Son's Salute

Jay Lacklen

From "*Flying the Line, an Air Force Pilot's Journey*"

During the summer of 1978, I got a chance to salute my mother with a B-52 flyby, of sorts, near her childhood home of Somerset, Kentucky. She had returned home to be with her mother in my grandmother's final days, and I happened to be flying a training mission on the Richmond Bomb Site with a route that circled Somerset. I thought a minor flight plan deviation would not ruffle any feathers and would provide a richly deserved tribute to my formative parent.

My mother had always harbored a lurking insecurity since marrying my father. She came from a small town and only had two years of classes at a nearby junior college, while my father had earned a master's degree from Stanford. It seemed she spent her entire married life trying to measure up by neutralizing her Kentucky accent, expanding her vocabulary, and pushing to get her kids accepted in ambitious social circles. Their marriage had failed some years before, and she feared that failure had somehow damaged her children.

Years later, I told her that if I had become a POW in Vietnam I'd have warned my captors they would have to flee for their lives when my mother found out they had me. She would have ripped the scraggly hairs off Ho Chi Minh's chin to get me released. She protected us as a mother bear protected her cubs and would throw herself between us and any force she felt threatened or endangered us. She didn't have a master's degree, but she had moxie and determination, some of which rubbed off on me. She deserved a salute from her firstborn, and I planned to give her one.

I told her I might be overflying and I'd have the local flight service station call her as we approached. When I contacted him, the flight service station controller seemed to get a kick out of arranging things, and air traffic control approved our slight diversion as we exited the route. Suddenly, flight service had hooked me up to the house phone and said, "Your mother is on the phone, Phantom Six-Zero, go ahead."

Dumbfounded, I could only think to blurt out, "Hi, Mom."

She asked how soon we would pass over and I said in about three minutes. She said she had to walk up the street to the top of the hill to see through the trees, but she would be there on time.

Three minutes later, we overflew the edge of town. We came in at 3,000 feet - high enough not to rattle windows but low enough to be clearly seen. I dipped the wings slowly left, then right. A few seconds later, I pushed up the throttles and the engines roared as I started the climb out of the route.

I like to think she felt some measure of awe, pride, and even wonder in remembering her little boy—who had once played with toy trucks in the corner by himself, been horribly shy, wet the bed, and stuttered. Now he had just flown over her in a giant aircraft to pay homage to her - the mother bear who had protected him, prepared him, and allowed him to soar.

Morning Alert Crew briefing

We Knew It Could Be a One-Way Mission

Tommy Towery

I was a senior in high school in 1964 when I attended a presentation on Buddhism at my church. While the religious theme of the night's program did not make a lasting impression on me, the person giving the presentation did. He was Japanese.

During his introduction he shared some information on his background and his words immediately caught my attention. He had been too young to join the military at the beginning of World War II, but he was in uniform at the end of the war. Had the war continued much longer he would not have been around to speak to us that night because of what he had volunteered to do in service to his country.

The speaker was learning to be a pilot at the end of the war and, had he completed his training, his death was assured. He was training to be a kamikaze pilot. To me, as a young 17-year-old high school senior living in a free country, the idea of giving up the rest of your life to die for a cause was unthinkable. He had been the age for having fun, and going to dances, and dating girls and planning for his future life. It was not an age when he should be learning how to die in a fireball by crashing his own aircraft into the deck of an enemy's ship. I did not think about how my father and the men with whom he served made a similar commitment in their own ways when they joined the Army. They knew there was always a chance of being killed in combat. Unlike the night's speaker, my father and his fellow soldiers also knew there was a chance they would survive the war. A kamikaze pilot knew when he took off it was a one way trip from which he would not return. He would either succeed with his suicide mission or he would be shot down and killed trying. It still sounds insane.

As ironic as it seems, I was unaware that not too far in my future I would be put into the same situation in its own strange way. Fresh out of college at the age of 22, I volunteered to make a career out of service to my country by serving in the Air Force, and doing so as a rated aircrew member. I was 24 by the time I finished my training to fly nuclear bombing missions as an Electronic Warfare Officer crew member in a B-52 bomber. In that capacity, in the event of a nuclear war, I would be asked to fly my assigned mission knowing my chances of survival were slim. Before doing so, I accepted the fact I and my crew would purposely make ourselves a priority target while living in an alert facility every third week. It was assured we would be a critical target of any first-strike nuclear attack against our country.

For the week of alert duty, crews were not confined to the alert building, but were not allowed to stray far from it and not leave the base under any circumstance unless a substitute crew member was first called in to replace him. Should that happen, the replacement had to be notified to report to base, study the mission, and assume alert with the existing crew, but only after materials were inventoried and signed for. During my years of alert, babies were born, family members died, and houses burned down and still the rules were enforced. Perhaps in the six years I was on alert a substitution happened one or two times, and still it took several hours for the replacement changeover to take place.

We were required to stay in radio contact with the command post and to be able to return to the facility at a moment's notice. We were constantly tested with alert exercises to insure we could respond in a timely manner. In the event of a real attack we would have less than 15 minutes to scramble to and climb aboard our assigned bombers and launch along with the rest of the Strategic Air Command's alert force. We would do so knowing once airborne we would be headed into a heavily defended enemy territory where our chances of survival were doubtful. Even if we survived the flight, the world to which we would return would never be the same as it was before we ran to the plane and took off on the mission. Looking back, I suppose in my own way I can now see that I too was in training for a suicide mission, but it was still not the same as dedicating my life to being a kamikaze pilot.

Fifty years later, as a man and not a boy, I can reflect on those days and see, and even somewhat understand, how a man is willing to die for the things in which he believes. It may still seem crazy to those who never saw a reason to wear a uniform and make such a commitment. It makes me wonder what a 17-year-old today would think if I told him I had trained to fly missions which would kill hundreds of thousands of people should it ever happen. That I sat on alert with my own and five other bomber crews for seven days and seven nights every three weeks for six years facing the possibility of doing so. We were confined to one building waiting for the horn to sound to send us scrambling to our aircraft and had only had 15 minutes from the time the alert horn sounded until being airborne and on our way to our targets.

Once airborne we would be required to fly halfway around the world and go into enemy territory at tree-top level in a massive eight-engine bomber flying at over 400 miles an hour. We knew we would be facing surface-to-air missile, anti-aircraft artillery, and jet fighter interceptors armed with air-to-air missiles and cannons en-route to our targets. Should we somehow be able to avoid the enemy's perimeter defenses we would have to bomb as many as four different targets before we could try to exit the enemy's homeland. We were briefed to expect each target to be heavily defended and our route timing was critical to allow us to even avoid the effects of our own nuclear blast. If everything went as planned, we would then have to land with our fuel tanks almost empty at an airstrip we had never seen before, if it had survived the hostilities.

Once back on the ground in a hopefully friendly nation, there was no guarantee we would ever get the support needed to get our aircraft refueled and airborne again, and even less guarantee we would have a home base to return to if we did. Aircrews were never trained to service their own aircraft, and without the assistance of a crew chief and a skilled ground maintenance crew there would be no way to re-launch our aircraft.

Today I look back and know I was very much like that speaker who I thought was crazy back in 1964, and know something else. In retrospect, both of us were willing to risk our lives for our countries, our freedom, and our families. By God's grace both of us were lucky we never really had to fulfill the mission for which we had dedicated our lives to doing.

I also know I was not alone in my dedication and commitment to that mission. The entire SAC alert force of bomber and tanker crews were just as well trained and dedicated; even knowing the chances of survival was slim. The ground crews and support personnel knew that once the aircraft were launched all they could do was wait to see if they would die in a nuclear blast. I have to wonder if I or any other SAC alert crew member were to speak to a group of high school seniors today, would they think we were fools for what we were willing to do to preserve their freedom? Would they believe we would actually fly the missions we were assigned and make the sacrifices we were expected to make? We trained for it, and if called, we would have done it. It was the Cold War, and we were the front line of defense. It was our mission. Peace was our profession, even if we had to sacrifice our own lives to preserve it.

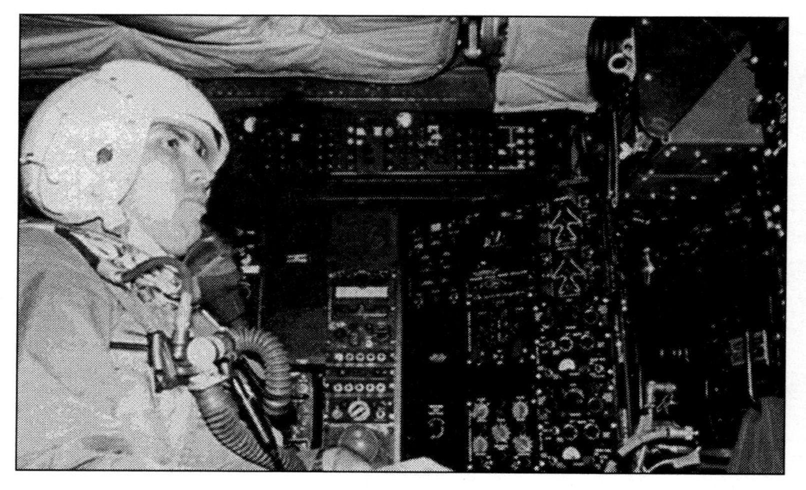

In my office

How I Got to be a Crewdog

Ted Lesher

Life is chaotic for everyone in a mathematical sense, in that we all are subject to small random events that in the end have great effects upon the lives we lead. A military career brings many experiences one would never encounter in the civilian world, and I have often been astonished when I reflect upon the unlikely chain of circumstances which put me in some particular life situations. This is my account of how some of those things came to pass.

For our 50th high school reunion my classmates and I were asked, among other things, to identify the person who had the greatest influence upon our lives. In my case it was an unemployed laborer I crossed paths with 55 years ago in an encounter which only lasted less than a minute. I have no idea who he was, and he never noticed me at all; but my entire life since then was guided by that incident.

First let me share some personal background. My father was a professor, a pilot, and an engineer who in the last two years of World War II worked at the Stinson aircraft factory near Detroit. I was five years old when I got my first airplane ride, in a Stinson L-5 with my father at the controls. I will never forget that first takeoff. When the ground fell away it was like the first drop on a roller coaster, scary and yet exhilarating at the same time. I suppose I have spent the rest of my life trying to re-create that magical moment.

When I was 16 I had a minimum-wage job delivering Western Union telegrams by bicycle for a dollar an hour. Between my junior and senior years in high school I used my savings to get my private pilot's license. I was flying a genuine yellow J-3 Piper Cub for four dollars an hour including fuel and paid two dollars an hour more for the instructor.

Then it was off to college, working my way through the University of Michigan engineering school with a job at a local record store where I was still making minimum wage. Tuition at the time was $50 a semester. I got halfway through my junior year when the Eisenhower recession struck. That was when the unemployed laborer came into my life. I was behind the record store counter when he entered and started looking through the records. Most of our customers were students or faculty, so he was a very unusual type patron for our business. I noticed him bump into another customer he evidently knew from high school and they had a brief conversation. That's when I overheard him say he worked in construction and was presently out of work. He remarked, "I can always go back to the Army – Uncle Sam always has a job for me." His words soon changed my life.

A few months later the recession caught up with the record store, putting it out of business, and putting me in a severe financial bind. I was ready for a change, so I quit school and hitched a ride across the country to Berkeley, California, where it turned out things weren't any better. I had some friends there who very graciously helped me out, but I was essentially homeless before they even had a word for it. I was from an academic family and there was an unspoken understanding I would continue that tradition, including working my way through college. I had never really considered anything else. Being in fairly desperate circumstances, I remembered the words of the construction worker and took a look at Uncle Sam.

I did some research at the Berkeley public library and discovered the Air Force had a program called Aviation Cadets offering both flight training and an officer's commission. It required applicants to have at least two years of college, which I had, and, given my interest in aviation and need for a job, it sounded too good to be true. I hurried to the local recruiting office and got signed up. They sent me to Travis AFB, California, where I went through three days of physicals, fitness tests, written exams, and psychological testing. I heard only a small fraction of applicants passed, but I was one of them. The first B-52 I ever saw was on the far side of the field at Travis (I actually thought it was a hanger until it taxied out and took off.) I never dreamed that in two years I would be crewed up in one and be closely associated with it for the next 35 years of my life.

The aviation cadet program was a vestige of WWII which trained tens of thousands of pilots and navigators, but the pilot part ended just before I entered. My eyesight prevented me from becoming a pilot so I ended up in the navigator training program. A year later about half of my cadet class had either flunked out or quit but I was nearing the end of training and doing fine. Then came the question of what I would do with my navigator rating. Aircraft assignments were based on class standing and I was going to have my choice. That's when another chance-event helped me make up my mind.

We had a number of academic classes and generally took one or two quizzes a week in each one. The tests usually consisted of about ten multiple-choice questions and I would rattle off the answers, hand in the test sheet and leave. After one test I was standing in the hall outside when one of the other guys said something about the questions on the back of the test. "Questions on the back?" We had never had questions on the back. I found the instructor and explained how I hadn't seen the questions on the back and needed to finish up the quiz. He was sympathetic to my plight but couldn't let me do it.

I busted the test and consequently had to post into a senior officer for counseling. He was a gruff and grizzled old navigator whose first words were "Are you trying to self-eliminate academically?"

I didn't know what he was talking about and sputtered something like, "No, sir. I just didn't see the back of the test." Once we had that straight and he saw the rest of my grades, he became fatherly and gave me some advice. He told me navigation was a dead-end career field and I should get myself into something better, namely electronic countermeasures.

And so it was. When the time came I chose to be an Electronic Warfare Officer and served as a crewdog EW for the next 20 years and 7,500 flying hours - all because I didn't see the back of a test. I'm not sure how good that advice was, but I think my life would have been considerably different if I had turned that test page over.

About a year later, I was nearing the end of EW school at Mather AFB, California, and reached another decision point in my life. There were various choices of assignments available for my class, but I really wanted to get into the B-52. Those were the LeMay years and the SAC bomber force was the absolute top tier of air crews and there was no reason not to go for it. Moreover, I liked California and was shooting for an assignment to the bomb wing across the base at Mather. You got to pick your assignments based upon your class standing and it came down to a fly-off between me and a close friend. We were essentially tied at the top of the heap, had one last in-air check ride remaining and the winner got the choice Mather assignment. Our airborne work was accomplished on converted C-54s that cruised up and down the coast north of San Francisco operating against a radar simulator site at Ukiah. We both had good rides but I edged him out enough to get assigned to B-52s at Mather, and to forevermore be a crewdog.

Official USAF Photo

Busted-Again

John York

This story relates to my experiences with our Armed Forces Security Forces: Military Police, Air Police/Security Police. Even though I have nothing but respect for these men and women; I have been unfortunate enough to eat vast quantities of concrete on too many occasions because of them.

I joined the Army immediately after graduating from high school, became an 11B (Light Weapons Infantryman), and carried an M-14 way too many miles. Although Army Military Police (MPs) were everywhere, most of the time they left people alone. Even when I was arrested for participating in a riot none of them pointed a gun at me as they marched about 50 of us off to the guard house. I really was not guilty of said charges; I just turned a corner and came upon a fight and stood and watched instead of running. I was cleared of any wrong doing within a day and went on with my normal duties. One night in the guard house was plenty for me.

After my stint in the Army I slowly worked my way through college. (What's wrong with taking six years to complete a four-year degree?) I managed to graduate at the height of our involvement in the

Vietnam War. Even though I was exempt from the draft I thought it was my patriotic duty to once again serve my country, so I took the written and physical exams for both the Navy and Air Force. While Navy flying seemed a little more exciting, I thought the Air Force was a better fit to my plans. Officer Training School was mostly a non-event and Medina Officer Training Base was uncontrolled. There were no fences, no gates, and no Air Police (AP). The year of 53 weeks (Undergraduate Pilot Training) at Laredo was also laid back and, except for an occasional AP driving by my BOQ room as we were partying pretty late into the night, I never saw them except at the main entrance gate.

My first post-UPT assignment was to Vietnam flying the highly subsonic O-1 Birddog. I lived in an Army compound and operated out of a 1,000-foot strip. We had ample security - at least I thought it was ample until we were overrun one dark night. In any case the Army MPs were always courteous and concentrated on the bad guys on the other side of the wire.

And then, and then, and then; along came Jones. My post SEA assignment was to SAC as a B-52 co-pilot. Unlike a number of my friends I did not consider that an insult. I hadn't planned on staying in the Air Force past the five-year point anyway. The first thing I noticed about the SAC base was that it seemed as though there were Security Police (the new handle for what had earlier been called Air Police) everywhere. I was blessed to have a wonderful Aircraft Command, Bob Reid, who took great care of me. (I even named our second son after him.) "Don't cross the red line except at the Entry Control Point (ECP)," he stressed. I'm sure I tested Bob's patience but I have always been grateful for the way he took care of me.

I did a quick in-unit upgrade to the left seat and even managed to stay out of trouble for a while. For a while! We deployed to Guam and U-Tapao along with all the remaining Ds and Gs as part of the Bullet Shot deployment. Things even rocked along several months before I was in trouble again. My crew lost a gun. Unforgiveable! Actually, it was an entire survival vest complete with the 38-special in it. No one seemed to care about the vest being missing, but the gun was a big, big deal.

We discovered we were missing the vest while we were en-route to a target in Laos. Upon return to Guam I immediately reported the missing gun and vest to the Security Police (SP), the Wing Command Post, the OSI, the crew bus driver and the short order cook who served us chili dogs at Gilligan's Island. A rather rude Master Sergeant SP came to our BOQ room to take statements from the crew. Two hours later he was still taking statements when I asked him to leave. We were tired (post-mission tired), and we were going to the Officers' Club to drink with the big boys. I fully expected to get nailed over the event and the TDY Wing Commander did offer me punishment under Article 15 of the UCMJ. I declined said Article 15 and my Squadron Commander (John Davis – a GREAT guy) got himself appointed as the investigating officer and he investigated the incident into oblivion. Luck was with me again.

Smith and Wesson Combat Masterpiece.

The dreaded "Rated-Supplement-Witch" crapped on my head and I spent four years out of SAC. It was not a bad gig: 0800 to 1600 daily, no weekend work, and I flew a T-39 at least one day a week. MPC was not inclined to allow this easy going lifestyle to last very long so I answered another SAC assignment, this time in the frozen North. As I returned to flying in SAC again I noticed two things: most of the crews were Kool-Aid crews and the SPs were meaner. On my first flight we started engines and taxied to the hammer-head awaiting our takeoff time. I asked the gunner very politely to pour me a cup of black coffee. He informed me that SAC crews didn't fly with coffee anymore; they drank Kool-Aid. Now, I had never flown any AF airplane without coffee and since we had a few minutes I asked the Supervisor of Flying to kindly swing by the flight kitchen and retrieve us a jug of coffee. Word quickly spread among the gunners that the new Major in the squadron was a mean old son of a bitch and he didn't fly without coffee, so forget that Kool-Aid stuff.

It was also obvious to me that there had been a change in the attitude/duties of the SP and they were pretty quick to point an M-16 at my head if they thought I was doing something wrong. I pulled alert every third week just like every other crewdog did. Our alert facility was on the opposite side of the runway from the base proper so we were always driving back and forth from the alert facility to the base/bomb squadron.

One morning the senior ranking AC announced that crews could no longer drive through the entrapment area of the alert facility without the full crew in the alert truck. Yeah, all six crew members. How dumb was that? Some of our guys had training scheduled in various parts of the base while others were scheduled to train inside the alert facility at the same time. We all growled about it but the Security Police held their ground. Their take was that if it took all six of us to fly the airplane then we needed all six to drive through the entrapment area.

One day I decided I had had enough of their stupidity and decided to press-to-test the policy. I took everyone with me but the gunner and drove up to the entrapment gate. The SP looked us over, opened the gate for us to enter and once we were inside closed the gate behind us. As we looked out the truck windows we could see at least 25 SP with their M-16s pointed at our heads. "Out of the truck and down on the concrete," was the command. So I'm eating concrete again. None of the other ACs saw fit to test the system after that.

This insanity went on for several months until one day we had a visiting General officer from SAC Headquarters on base and, as usual, he wanted to visit with the alert crews. He came in with the Wing King and some other O-6s and quickly began to ask what our lifestyle was like and how we were being taken care of/treated. I was rapidly on my feet and spoke of the entrapment insanity. He listened, quickly turned to the Wing King and asked him "What the Hell?" The King seemed to plead that he had no knowledge of the situation, which made things even worse. By end of day we were allowed to drive the alert vehicle through the entrapment area with six people or one people or a shitload of people in the bed. Score one for the good guys.

One day I was relaxing in the alert facility when my crew chief came to me and said he needed the maintenance log out of the airplane. We went through the Entry Control Point (ECP) and straight to the red line surrounding the airplane where the close-in SP met us and checked credentials again. Now I had thought this through. I was well acquainted with the two-officer policy regarding nuclear weapons. The maintenance log was kept at the foot of the stairs immediately next to the entry hatch. I knew I was not going to enter the airplane, but just reach inside the hatch and retrieve said forms. I had done the very same thing many times before. Well, I no more had unlatched the hatch when I heard this SP radio blaring "Two-officer policy violation! Two-officer policy violation!" That time there were at least 25 fine, young SP facing me and every one of them had a gun pointed at my head. "On the ground, York! On the ground!" "Eat some more concrete York. Eat some more concrete!" A senior NCO SP quickly arrived to assess the situation. The young SP had a different spin on things but the NCO quickly straightened him out and the Crew Chief and I returned to the alert facility complete with the maintenance log.

Here's the last part of my story. I was on alert as usual and my entire crew was on the main base and we responded to a klaxon. We drove to the main runway and tower gave us a green light to cross. The gunner was driving, as usual, and we drove straight up the throat to our alert bird. As we entered the throat there was an SP there who challenged us with a certain number of fingers. We returned the challenge with the appropriate number of fingers and drove on to our airplane. I was the first one up the ladder, as usual, quickly followed by the co-pilot. Soon we had all engines running and equipment on line. The Emergency Action Message was decoded as a Polka Dot something or another which directed us to record our ready-to-taxi time then shut down our engines. Before we started the shutdown procedure I noticed a lot of activity around our airplane. I asked the co-pilot if he was seeing what I was seeing. There were at least 40 or 50 SPs standing all around the red circle and they were all pointing their M-16s at my head. Not again, I thought; this can't be happening again.

At that point I decided I was not going to willingly eat any more concrete again and I left the engines running. The exercise checklist called for us to shut the engines but I did not. The Wing King in his white-top car pulled up front of our aircraft and he told us to shut down our engines; and I did not. I knew that there wasn't a single person who could come inside that circle without my permission. That and the fact that we had a bomb bay full of … well, I can neither confirm nor deny that there were nuclear weapons involved. I finally shut down the engines but we did not leave the airplane. It was a Mexican stand-off. The Wing King, the DO, and I don't remember who else were involved but I was not going to eat concrete again, especially when I had the advantage for a change. The DO directed me to come out of the airplane so the SP could jack me up. I told him I didn't think so and asked him what this was all about. He replied the SP at the throat thought we held up the wrong number of fingers in response to his challenge, but he wasn't entirely sure. I told him that we were not deplaning until the SP leadership was involved and straightened out the matter. I also told him that we were not to be jacked up and I was not eating concrete again. The O-6s huddled, more SP joined them but at least the latter ones weren't pointing their guns at me. Very quickly the mass of young folks brought their weapons down and Wing King informed me we could come out without fear; that we would not have to eat concrete. I had been in SAC long enough not to believe what he was telling me but all the crew heard the transmission and were willing to deplane. We did so, did the normal post-engine-start procedures and drove our alert vehicle back to the facility.

Naturally I was in the Squadron CC's office in 10 minutes to get my AC butt chewing but I smiled throughout it because, just for once, I knew I had held the upper hand.

Eventually I moved on to another SAC base and after a couple of years of flying and pulling alert I wound up in a Deputy Dog position where I wasn't in charge of anything but took the blame for any and everything that went wrong. I tried to go by the alert facility at least three or four times a week just to see how the guys were doing and to listen to their bitches. I could not do anything about their bitches but I was a good listener. On one particular day the senior AC sought me out and told me his crew had a brand new gunner. (Brand new - his first alert tour.) Since there was no ground training that afternoon he thought I might want to put the new gunner on ORI watch or maybe thunderstorm watch. Well I played along and I took the new gunner out to the flight line and stationed him in a certain place and told him that his job was to watch out for thunderstorms. If one was to come near the base he was to jump up and down and clap and get every ones attention about said thunderstorm. Well, I got called in to a meeting I hadn't anticipated and it lasted longer than it should have. When I walked out of the hanger to get back in to the staff car my heart stopped. Yep, a thunderstorm had appeared and the young kid was still standing where I left him, drenched to the bone. I really felt bad about that. It was not funny and I did lots of apologies to him, his AC, Squadron CC and anybody else that would listen.

I will leave you with my thoughts about our Armed Forces Security Forces. They are top notch men and women doing a very difficult job and I salute every one of them.

Sworn in over North Vietnam
Mike Jones

The day of 12 March 1969 was a very special day for me at Andersen AFB, Guam. I was scheduled to fly what was probably my last B-52 combat mission. I was there TDY from Ramey AFB, Puerto Rico, and my six-month TDY tour was about to come to an end. I had just found out I had made Technical Sergeant, and didn't even know I had a line number. Communications from Guam to Puerto Rico was difficult, with a three-minute phone call costing $45.00. Those calls were frustrating and full of static and echoes since they traveled via an undersea cable to California, then across the states on a land line to another undersea cable to Puerto Rico. A letter from an FPO such as Andersen to an APO like Ramey took more than a week in transit, and a response another week to get to its destination. I was a TSgt and didn't know it for several days before the message from Ramey finally got to me.

The Crew Chiefs on Ramey were given a flying slot every month. When I got my TDY orders to Andersen, I was given six months of flying slots to allow me to fly while I was there.

The day I am talking about was special since it was the day I was due to reenlist. After the crew briefing I asked the Aircraft Commander if he would re-enlist me during the mission and he agreed. Our aircraft was 55-0051 and we were flying as Tan 3, the third aircraft of the three-ship Tan Cell. The flight crew was crew E-12 from the 454th at Columbus Mississippi, with Capt Overstreet as the Aircraft Commander.

The cell took off on time and everything was routine including the air refueling leg. When we were 50 miles off of the Vietnam coast I passed out the crew's flak equipment. The Guard frequency was full of war transmissions, artillery warnings, Jolly green extraction calls, and strike reports. It always brought home the reality of where we were going and what we were doing. It was all business from then on.

As we reached the MiG Threat line 50 miles south of the DMZ two F-4s appeared off of the number two aircraft's wings. Our fighter escort had arrived. That was my second time crossing the DMZ which looked like the moon to me. There were water filled bomb craters and nothing green for miles. We were flying slightly west of the DMZ in Laos when the radio announced, "Tan Cell, angels unfriendly 11:00 your position."

I was sitting in the IP seat and had a good forward view. We all strained to spot aircraft approaching the cell. Off in the distance, we spotted the trail of black smoke of a MiG-21 in a wide turn approaching the cell about five miles out. The two F-4s banked hard right and set off on a course to intercept the MiG. Just after it passed a few miles in front of the lead aircraft, the MiG did an immelmann maneuver and passed within a mile off the nose of the cell's lead aircraft. The F-4s were way off to our right as the MiG headed back from where it came. So much for fighter protection.

Shortly afterwards we turned in-country and began the bomb run. As the cell released the bombs I watched for secondary explosions but saw none, which was common from the altitude we were flying. As we headed back the same way we came, Capt Overstreet read the oath of enlistment over the intercom. We were still over the north.

The flight back to Guam was uneventful, as usual, but I learned years later, Crew Chiefs and other maintenance troops were not allowed to fly if the aircraft was going North. In March 1969, that was not the case.

It was a unique experience.

Goin' to School While Goin' to War

Randy Wooten

By the spring of 1973 I was a high-time B-52D co-pilot with a tour as a forward air controller (FAC) in Vietnam under my belt and was TDY again to Guam and Thailand. The war was still going on, and it looked like an end was still nowhere in sight. I had accumulated over 1,600 hours total flight time, and nearly 1,000 hours of combat time.

SAC had a big problem - they didn't have the resources to pull crewmembers out of the TDY rotations to begin upgrades from co-pilot to aircraft commander, or nav to radar nav. So they decided to do "local" upgrade rather that spring the guys lose to send them to the real Combat Crew Training School (CCTS).

I was TDY to Guam as the co-pilot on a crew, and through distribution, received a form informing me I had been selected for local upgrade to "pilot." Being the war-weary, smart-ass that I was, I replied I already had my pilot wings, and had been functioning as such for over three years. I stated I would however accept their offer to upgrade to aircraft commander. And so it began.

Today I look back over my flight records, and shake my head in amusement. On any given week I would have a simulator ride, fly a combat mission, and also fly a training sortie. This short story is about one such combat/evaluation mission.

As you can imagine, KC-135 tankers were in short supply at Guam. The 12-hour combat missions to Southeast Asia took more than one tanker per B-52D that was carrying over 60,000 of weapons which were loaded in the cavernous bomb bay as well as hanging on the wings. So the obvious solution was to put upgrading co-pilots on combat sorties with instructor pilots on the crews.

To me, refueling was voodoo. I'd tried my hand at it as a co-pilot along the way, and the first few seconds went pretty well. Then I'd start oscillations behind the tanker that got larger and larger, followed by the instructor saying, "I've got it." I was so damn frustrated, telling myself that if "they" (meaning all the co-pilots before me that became ACs) could do it, so could I. Well. I was beginning to doubt it. I've decided the real heroes of the war were the crew members who were strapped into their ejection seats during my many attempts to fly the plane into the small box called the air refueling envelope.

I finally got the hang of it and the training guys put me up for my check ride. But it wasn't the typical check ride I expected. It looked a lot like the training rides, which had me sometimes fly along with other crews to accomplish a specific requirement, like a night landing. This approach was not at all like SACR 60-4 (Strategic Air Command's Aircrew Training Regulation) strongly suggested. The typical training and check ride sortie had the student act as the aircraft commander for the entire sortie - mission plan, takeoff, air refuel, go low level, fly a celestial navigation leg, return to the takeoff base, and then do pattern work. That typically took eight hours or so.

But, after many sorties, I had all my requirements accomplished. Then began the series of sorties with pieces of check rides embedded. My most memorable one was to accomplish the refueling check - a 15-minute contact with no more than three inadvertent disconnects. And, oh by the way, it was to be accomplished at night, with a heavy-weight refueling.

I can still remember being somewhere over the Pacific, at 0100 with 108 500-pound bombs, trying to take on 100,000 pounds of fuel from two tankers. It was in a three-bomber formation with four tankers, so each bomber had to take fuel from two different tankers. I did it, but it wasn't pretty, and I was completely fried after my 30-minutes in the left seat. I know the guy who took over after me wondered how the parachute back-pad got so wet.

So technically, I was officially an aircraft commander, but I was still needed as the crew's co-pilot. About that time the war ended and most of us came home. Upon arrival back at my home base I was informed that my upgrade training on Guam didn't count and I would have to start all over again per the more structured SACR 60-4 fashion.

I then did a "real" local check-out at Dyess as the co-pilot on a stan/eval crew, got my "real" SACR 60-4 check, and got my own crew. I was an aircraft commander for 11 months, and took my own crew back to Guam and Thailand. When I returned to Dyess I promptly entered the local IP program - so it was back to the right seat.

Someone once told me that as a pilot in the Air Force, you'll spend very little of your time actually doing your job. Instead you'll spend most of it in the school house, as either a student or instructor. I believe it.

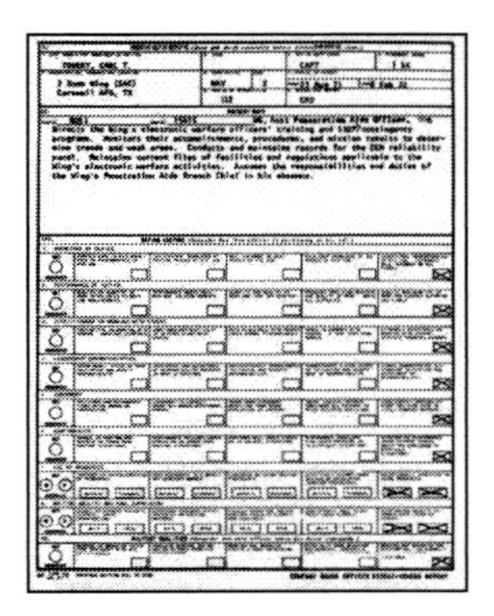

The Career Ending "Promotable Three"

Tommy Towery

While not specifically a problem unique to just B-52 crews, I believe the "Controlled" Officer's Effectiveness Report (OER) system of the 1970s did more to deflate crew moral and empty the B-52 crew force of highly qualified officers than any other factor during that period.

According to Air Force Regulation 36-10: "*The purpose of the officer evaluation system is to provide the Air Force with information on the performance and potential of officers for use in making personnel management decisions, such as promotions, assignments, augmentations, school selections, and separations. It is also intended to provide individual officers information on their performance and potential as viewed by their evaluators.*"

When I entered the B-52 crew force in 1972, each officer crew member was evaluated by the officer who supervised him. At the time there were no female B-52 crew members. In the case of most B-52 crews, the rating officer was normally the Aircraft Commander of the crew. Rare exceptions happened when the Radar Navigator or the EWO on a crew outranked the Aircraft Commander.

AFR 36-10 listed all of the events requiring completion of an OER, but the most common were: *a PCS move by the rater or ratee, or a change of assignment. As a minimum, an OER had to be completed at least every six months for lieutenants with less than three years of service, and annually for all other officers through colonel.*

From June through September of 1987, Syllogistics, Inc., in conjunction with the Hay Group, conducted a study to examine the strengths and weaknesses of the then-current United States Air Force Officer Evaluation Report (OER) and to recommend alternative designs which could improve its usefulness. Their study included an examination of past OER systems, including the Controlled OER era.

In the report they noted the rater was responsible for collecting all the additional information he needed to complete an OER. Typically, the rater would ask the ratee to provide an update on his/her accomplishments during the rating period, and solicit information on the ratee's performance from other supervisors who had observed the ratee's work. The rater completed the rater portions of OER, and then submitted it to the additional rater for completion of the next portion. The additional rater added comments, signed the form, and forwarded it to the endorser for final comments and signature. The endorser returned it to the CBPO for further processing and quality control in most cases.

At the time of the Syllogistics study in 1987 there had been six distinct phases in the Air Force OER system since the establishment of the Air Force as a separate service in 1947. These were:

1) The forced choice method 1-1 adopted from the Army in 1947-49.

2) The critical incident method used from 1949-52.

3) Rating of performance factors with narrative commentary, 1952-1960.

4) The "9-4" system, 1960-1974.

5) The "controlled era", 1974-1978.

6) A return to a mechanism similar to 3) from 1978 to the present (1987).

During my time as a B-52 crew member I only fell under the requirements of the 9-4 system used by the Air Force up to 1974 and the controlled OER system from 1974-1978, so those are the ones which directly impacted my career.

The Syllogistics study made many other observations: (Extracts from their 1987 study will subsequently be noted in *italics.*)

The primary purpose of performance appraisals in the private sector is to make short-term compensation-focused decisions. An OER in the Air Force has far-reaching promotion and career implications for the individual officer.

The Air Force like many large organizations has experienced inflated evaluation ratings and/or evaluation systems which were incompatible with their overall purposes.

Two characteristics have recurred throughout this history.

In 1960 the "9-4" system was begun. The 9-4 system continued to use the overall 9 point scale evaluation from previous systems but added to it a requirement to rate promotion potential on a scale from 1 to 4. Initially, the 9-4 system did bring some discipline to the ratings but eventually the ratings became "firewalled" at the top score of 9-4. This inflation occurred even with an extensive educational program to warn evaluators against rating inflation.

By 1968 ratings inflation had once again rendered the OER system ineffective. Nine out of ten officers received the highest rating, 9-4. Development work on a new system began in 1968 and continued through 1974 when the controlled OER came into being.

My personal OERs as a Captain Electronic Warfare Officer (EWO or EW) during these early rating periods were indeed 9-4 ratings and, as such, I believe kept me competitive for future promotions. However, that was soon to change.

In 1974 the controlled OER era began. The basic form of the previous OER was retained but raters were instructed to distribute their ratings as follows: 50% in the 1st and 2nd blocks (two highest) with a limit of 22% in the highest block. Although the system had been extensively discussed and pretested prior to implementation, it encountered almost immediate resistance.

Carswell AFB, Texas, where I was stationed at the time of this change in rating systems was home to two B-52 squadrons, the 9th and the 20th. The data contained on recall rosters during the controlled era for the 9th Bomb Squadron in 1977 shows there were 66 captains assigned to the crew force, of which 21 were designated as Aircraft Commanders. The 20th Bomb Squadron had 79 captains with 24 of them serving as Aircraft Commanders. In fact, all 17 crews in the 20th and 18 crews in the 9th had captain Aircraft Commanders.

If one does the math, it will show only 14 of the 66 captains in the 9th could receive the highest (top 22%), rating and only 17 captains in the 20th. That means the new rating guidelines did not even allow every Aircraft Commander the capability of receiving the highest OER evaluation rating. These limitations had a much more severe impact on the other 114 Radar Navigators, Navigators, and EWO captains in the two B-52 squadrons. The feeling at the time in SAC was that the most important person on a bombing crew was the Aircraft Commander (Pilot) followed by the Radar Navigator who was responsible for getting the bombs on target. There was also an overall feeling among the crew force that the Air Force paid special attention to the career track of Air Force Academy graduates.

The basic problem with the controlled OER was that officers who were experienced in a system that gave top marks on just about all evaluations understandably resisted a system where top marks became the exception. Perceptions centered about the notion that a "3" rating was the end of an upward career track in the Air Force.

Although educational efforts were made to overcome such misgivings and ultimately only the top block was controlled, the initial anxiety about the system was never overcome.

The controlled OER (1974-1978) struck directly at the inflation problem by requiring a forced distribution of ratings. The perception at that time was that a 3 rating or below was akin to the end of an upward Air Force career track.

In my own case, after receiving my first controlled three I knew I needed to attempt to find a better way to stand out from the crowd of my fellow EWOs. I believe all the captains in my career specialty also received the same controlled three rating. My rating came despite being an Instructor Electronic Warfare Officer (IEW), an outstanding performance rating for my crew on the previous Operational Readiness Inspection (ORI), and my crew being recognized as a Blue Ribbon Crew during the reporting period.

My plan of action was to elevate myself to a position where I had a greater chance of receiving a higher OER by becoming so proficient in my duties I would earn a position on one of the wing's elite Standardization and Evaluation crews. Those crews were, by definition, the highest qualified crews in the unit. My feeling was, if I was the best EW in the wing I would be recognized and rewarded.

Many officers feel that since they cannot stand out on the basis of their ratings they must pursue certain types of education and assignments, which may have nothing to do with preparing them to assume greater responsibility, in order to provide the promotion board with the proper "image". A corollary to this phenomenon is the feeling of unfairness caused by the fact that certain primary assignments make it much more difficult to accomplish these peripheral activities. For instance, certain aircrew members may find it impossible to attend evening classes to improve their educational attainments on a regular basis, if much of the time they are away on temporary duty (TDY).

I volunteered to head and reached 100% participation as the squadron's Combined Federal Campaign program; a hated job which no one wanted. I became a multiple contributor of the wing's suggestion program and received several awards for suggestions on aircraft modifications and crew procedures which were approved. I volunteered for any job I felt could made me stand out from the crowd. I even took classes three nights a week and earned my Master's degree from the educational program began by Texas Christian University through an on-base program. I don't know exactly how I accomplished the goal I set, but in the end I did, and by the time my next reporting cycle came due I was serving as the EWO on Standboard crew S-01, the most elite crew in the squadron.

It turned out all my efforts were in vain and when I reviewed my next OER, I was shocked and disappointed to find I had once again received a controlled three. Even after all my hard work, I ended up with the same ranking as any new EWOs who had just signed into the squadrons and had done none of the additional duties I had performed.

It was then I realized my only way to have a chance of ever receiving an OER with the potential to get me promoted was to get out of the B-52 crew force entirely. While some had made their way to staff jobs, all of the EWO staff positions were filled by senior ranking officers (some just awaiting retirement.) It became clear my best chance of receiving any potential promotable OER was to find a position in a unit where the EWOs did jobs more important than defending an aircraft during peacetime training sorties. For me and many other B-52 EWOs, I looked to the RC-135 program, with an aerial reconnaissance role with national implementations. After several lengthy sessions with the assignment section of the Military Personnel Center, I was assigned to Offutt AFB, Nebraska, where I joined a squadron of highly professional EWOs, most of whom were previous B-52 crewdogs. The reconnaissance missions were tasked by national intelligence agencies and were still considered high-threat missions resulting in Air Medals awarded even though technically they were still peacetime missions.

Alas, the new mission and the accumulated Air Medals did not offset the damage already done during the controlled three OER periods. This became obvious to me when I found out that I, along with all seven other former B-52 captain EWOs currently in the RC-135 squadron, failed to be selected for promotion to Major.

Because of budget requirements, legislative controls and a number of other factors, the Air Force system requires an officer either to be promoted at each opportunity or to leave the service at some point prior to completion of a full career. It is this fact that places so much of a burden on the OER system. There is no parallel in private industry whereby one performance appraisal can, in effect, dictate a decision to lay off a person many years in the future.

For most of the group a second non-selection the next year brought an end to their service to our country. At that time in the Air Force system, two passovers for promotion resulted in automatic separation from the service.

This is where fate played a strange card in my life. The resulting exodus of highly qualified EWOs created a critical shortage in my career field. To remedy this, the Air Force Personnel Center implemented a program which offered certain highly qualified officers in recognized Critical Career Fields an opportunity to remain in service even though they had been twice passed over for promotion. The extension was for a two-year period during which the officers would still be considered for promotion, though few actually were promoted. At the end of the two-year period these officers could once again be extended for another two years should their career field remain critical, and the process continued until they were eligible for retirement. I was not valuable enough to be promoted, but too valuable to be kicked out.

For me personally, this continuation had a silver lining I never expected. I had never been augmented into the Regular Air Force and was still serving as a Reserve Officer. By being selected for continuation, I remained on active duty long enough to be considered for promotion by the Reserve Promotion Board.

The past practices of the "two passovers and you are out" resulted in most Reserve Officers being separated before the Reserve Promotion Board met at the 14-years-in-service point. Aided by my service record of B-52 combat time, RC-135 reconnaissance missions, and the resulting Air Medals and Meritorious Service Medals, I was selected for promotion by the Reserve Board. There was one catch - I remained on active duty as a captain even though I was officially a Reserve major. Thus, I wore my captain bars for 17 of the 20 years of my service, and on the day I retired, I pinned on my gold oak leafs and today draw the retirement pay of a major.

I suppose in my own way I beat the system.

These are some of the final observations made in the Syllogistics study:

In 1978 the controlled OER era ended when the Air Force leadership decided that individual need for a less stressful OER system was more important than the management benefits of differentiation.

The controlled OER generated a great deal of anxiety and loss of morale which are well remembered today.

Our interviews and focus groups indicated that the controlled system has left deep scars within the officer ranks. It has an almost uniformly negative image and people are quick to relate instances of "good" officers leaving or being forced out of the service because of a "three" rating.

A Young Ken

Destined to Be a Crewdog

Ken Charpie

I can't tell the tale of my days as a young B-52 crewdog without beginning the story a little earlier in my career. I started out intending to be a pilot as a member of Class 80-01 at Vance AFB in November of 1978. Around February of 1979, I washed out, but due to a medical issue on the part of the Wing Commander, had to wait until May before

I could finally get orders from Vance to Mather AFB for Nav School. I spent that time on casual status, and had to organize the graduation ceremony for a class from Vance in April. Gen Bennie Davis had just become the commander of Air Training Command, and it was his first UPT graduation to attend as the Commander, and I was there.

In May 1979, I finally went to Mather as a member of Class 80-03 at UNT. Come time for graduation, who should be the VIP guest to talk to us? None other than Gen Bennie Davis attending his first UNT graduation as the Commander of ATC, and I was there. First UPT graduation in April; first UNT graduation in November. Tells you all you need to know about pilots and navs in the Air Force.

After spending time at Advanced Nav Training, Water Survival, Nav-Bombardier Training, then B-52 Combat Crew Training at Castle AFB near Merced, I finally arrived at Minot AFB, North Dakota, in October of 1980.

I went through the usual training to get ready to pull alert, got certified for the Emergency War Order (EWO) mission, and went on alert for the first time. At the morning briefing, the Wing Commander came in to have a few words with the crew force. At the end of the briefing, he looked at my AC, and said "John, after preflight, bring your crew to the alert pad library. I've got a special job for you." We went out, preflighted the plane and weapons, and went to the library room as requested. In walked the Wing Commander with Bob Schieffer from CBS News and a camera man. Our special job was to be interviewed for a "Defense of the Nation" special for CBS. The first question Bob Schieffer had for us was, "What do you do on alert?" All I could think of was, "As far as I know, we get interviewed by CBS News every tour." I had a whopping three hours of alert time under my belt by then. Our interview ended up on the cutting room floor. My father-in-law was a news anchor for a CBS station in San Diego at the time and later I asked him if he had anything to do with the choice of base for the special, but he denied it.

Gen Bennie Davis went from ATC Commander to Commander in Chief, Strategic Air Command, and where did he go for his first visit to a B-52 alert facility as CINCSAC? Minot AFB. And I was there.

My crew was selected to participate in an exercise flying out of Biggs Army Airfield near El Paso. At Biggs, we aircrew types got to stay in the bachelor quarters on Fort Bliss instead of staying in "tent city" at Biggs, so we could get crew rest in air conditioned comfort. One afternoon, my radar nav and I were watching MASH on the cable CBS station. At a commercial break, they went over to a quick news update. I wasn't paying much attention, but then it hit me. It was my father-in-law on the TV! The El Paso cable system had a San Diego CBS station channel. That was the first, and only, time I ever saw my father-in-law live on TV.

One winter, we were having a real hard cold spell up at Minot, and I got a call at home from my father-in-law. He started asking all sorts of questions about the weather, and how we handled living in those conditions. A couple weeks later, we got a package in the mail with a tape of the conversation, and a little more. It turned out, my father-in-law was on the air doing the evening news and during the weather forecast, and the weatherman contrasted the nice weather San Diego was having with the cold spell "up north." My father-in-law had picked up the phone ON AIR, and called us up for a "first hand report" of the cold weather!

One day, I had to go visit the Ops Officer for something; I don't remember why. He had a board behind his desk with the names of all the inbound personnel by crew position. I noticed under AC, there was a Major Milberg listed. I asked, "Is that a Major Raymond S. Milberg?" He shuffled some papers, and replied, yes, it was, then asked how I knew him. My reply: "He was the flight commander who washed me out of UPT." "Good, then you already know your new Aircraft Commander!"

At the time, SAC had a three-year policy for northern tier bases such as Minot. Spend three years in the cold, and they'd theoretically work with you for a southern tour. When the SAC Assignment team called me around the three-year point to see what my preferences were, I told them I was happy where I was, and wanted to just stay at Minot. They offered me an inducement - a class assignment at the Central Flight Instructor Course (CFIC) en-route to a base on my Form 90 (aka Dream Sheet), KI Sawyer. First, I knew the Combat Evaluation Group had just gone to KI and busted nearly every instructor on base. I had no desire to be there in what could prove to be a bad time rebuilding the unit, and second, Minot already had a CFIC spot for me, in the very same class. When I told the assignment officer that, he replied he didn't show me in that class. I told him "You've got Phil Krasnicki in it, right?" When he confirmed it, I told him, he wasn't going, I was. Phil had a later slot, after he first went to Bomb Comp. I stayed at Minot for a total of almost five years.

My crew was selected to participate in an exercise that would have us take B-52s into Ramey Air Force Base in Puerto Rico. Unfortunately, as the planning proceeded for the exercise, it was learned Puerto Rican nationalists threatened to disrupt the exercise. The planning was changed to have us deploy to Homestead AFB in Florida instead. As we took off in late April, a blizzard was moving in and

Minot got two feet of snow while we were on our way to Florida. That was an unusual occurrence for Minot, as normally the snow was horizontal, and just passed through on its way from Montana where it started falling to Minnesota where it landed. The wives were not amused at having to clear the snow while we "relaxed" in the South. Actually, the snow melted just as fast as they could shovel it.

The scheduler promised every crew involved a chance to hit the beach or town for one day with no duties. However, on the day my crew was scheduled to be off, we got called in to the planning shop. We were asked to plan out a flight to Ramey, land, spend a little while there, then take back off and return to Homestead. It was kind of thumbing our noses at those who kept us from going into Ramey in the first place. The leadership wasn't completely satisfied with our first plan, so they kept us busy all day long planning alternative scenarios for the diversion. At the end of the day, which option got flown? Yep, our very first one. My crew never did make it to the beach.

After about four years, I started working to try to get an assignment teaching Chemistry at the Air Force Academy. I went out to the Academy on my next leave period to interview with the department. When I got to the first meeting, they pulled out a piece of paper, and said "Ken, sorry to have to show you this, but we just got this message from Military Personnel Center. It said: "B-52 navigators are critically manned and unavailable for assignment to the Academy until further notice." They went ahead with the interviews to keep them on file for future reference if needed.

Not long later, I got the opportunity to go to Squadron Officer School at Montgomery AFB. One of the things that happens there is the MPC assignment people come out to talk to my class about future assignment possibilities. When the officer handed me the obligatory one sheet summary of my career, I noticed it had some handwriting on it. "Congratulations on your AFIT assignment!" So, I asked if I was going for my Masters Degree in Chemistry so I could go teach at the Academy. "No, B-52 navs are critically manned, so you can't do that. Instead, you're going to get a Masters in Strategic and Tactical Sciences because we have to have rated people get that." I couldn't get one Masters Degree because I was a nav, but I could get a different one for the same reason. Oh well, that allowed me to do something other than fly, and I got to work on the operational testing of the B-1, and B-2, as well as a number of other programs.

While working on the B-1 testing program at HQ Air Force Operational Test and Evaluation Center, I did not know how to spell B-2. But, at some point, a special test office in the Center brought two people to talk to me. They wore civvies, could only tell me they were "John" and "Tom", and nothing about what they did. They had lots of questions about the B-1 and our testing, especially testing the simulator system. I was assured I could talk to them freely.

After four and a half years at AFOTEC, it was time to go back to flying. I filled out my new dream sheet for the upcoming assignment, volunteering to go back to Minot AFB. I hadn't heard anything from the assignment folks in a while, so I gave them a call to see what they were planning for me. The assignment officer started off by asking me if I'd like to cross train to the B-1. I was a junior major by then, but had no desire to be the new guy in a new weapon system, so I said "no." OK, what southern base would you like? With that question, I knew he had my record right in front of him, to know I had already spent five years up north and was due a southern assignment if I desired. So, I asked him to see what he could do for me that was actually on my dream sheet. "OK, we're sending you to Minot." That was fine with me, and I talked to my wife after work. She said, she'd been thinking. Uh, oh. And, she'd prefer to go to KI Sawyer.

I filled out a new Form 90, and gave it a little time to percolate in the system, then gave the assignment office a call again. "How would you like to cross train to the B-1?" My reply, "Haven't we had this discussion before?" They sent me to KI Sawyer. My assignment was designated from the start to be a Flight Commander in the Bomb Squadron.

After we got to KI, I was assigned my flight of six crews, and I had to go back through the certification process so I could start pulling alert with my crew, and with my flight. One part of the training called for me to have an alert pad tour. Five years of having pulled alert at Minot, and I was still required to have an alert pad tour. My flight was going on alert on Thursday, so we scheduled the tour to be Monday morning, and I would get the tour and a good opportunity to meet my flight at the same time. However, over the weekend, I was watching the news, and there was a special presidential announcement broadcast. During the announcement, President G.H.W. Bush proceeded to cancel every program I had been working on back at HQ AFOTEC that was still in test or planning, so my replacement there suddenly was out of work. He then proceeded to take my flight off alert! That was something for which we did not have a plan. We never did find out how the bottle of champagne got into the alert facility.

While I was on a crew, my AC's sister was working for Congressman Bustamante out in DC. When it came time for the Andrews AFB Air Show, she got the congressman to send a request to SAC for crew E-75 at KI Sawyer to bring a B-52 into the air show. So, my crew got a "request" to come see the Wing Commander. "Why do I have a Congressional request for YOU to go to the Andrews Air Show?" Well, with a Congressional request, there was not a lot to be said or done about it, so we flew to Andrews. As we were taxiing into parking, we were going head to head with the Blue Angels, when we suddenly had brake failure. The pilots said the Blue Angels did a bomb burst pattern on the ground when they realized we could not stop. We had a good time talking to people about the B-52; the H-models were only a little over 30 years old at the time, but one spectator was asking such directed questions to warrant us to go talk to the Office of Special Investigation (OSI) about him upon our return to KI.

Our return to KI - that's another discussion! To fix the brakes, they shipped in some new brakes and a specialist to fix them. However, as he looked at the new parts, he decided they were defective. More parts, more problems. The specialist talked to his office, and they figured he could build up good brakes from pieces and parts from all the bad brakes. It took a couple of extra days, but finally we were able to fly home.

When we turned in our rental car, we found a bullet hole in it. I don't think they had put us up at the best possible local hotel.

Andrews AFB was not built for B-52s with the wide wingspan, and tip gear hanging down. The taxiways were too narrow. It had not been a problem on landing because of the light weight at the end of a flight, but taxiing out coming home with more fuel onboard became a problem. The pilots had to keep shifting fuel from side to side to lift one tip gear, then the other to avoid taking out all the taxiway lights! I think we still took out four of the lights, but were congratulated for taking out fewer than any other previous B-52 crew had. We finally got to the end of the runway, and were cleared for immediate takeoff. So, we taxied into position, and – "Well, sorry, but we have to sit here and shift fuel back where it belongs to have weight and balance right for flying."

All of a sudden an aircraft called in on 10-mile final - Air Force One. "Havoc 75, cleared for takeoff. NOW." As Air Force One kept calling shorter final, the tower controller's voice kept going up. We finally started takeoff roll with them on about two-mile final. I think everyone was glad to see Havoc 75 leave Andrews!

We got to re-organize the wing with the transfer from SAC to Air Combat Command. I moved up into what had been the Deputy for Operations staff, but was then called the Operations Support Squadron. Then came the next Defense Base Closure and Realignment (BRAC) plan, and KI came out on the closure list. I was in a position where I got to brief visiting dignitaries on the base and wing mission in an attempt to persuade them not to close the base. I briefed the Secretary of the Air Force, a Vice President for RAND Corporation and head of Project Air Force for RAND. It was all to no avail; KI was confirmed for closure. So, it was again time to look for a new assignment. The Air Attaché to Denmark position didn't work out for me, but I was able to find a job working on the operational test team for the B-2 at Edwards AFB.

When I got briefed in to the B-2 program, who were the first two people I met? Lt Colonels Harris and Cressman: "John" and "Tom" from all those years before! Both were also former B-52 crewdogs!

I closed out my Air Force career working in the B-2 Requirements Office for Air Combat Command, but working in the B-2 Program Office at Wright-Patterson AFB. So, what am I doing in retirement? I teach part time at Wright State University near Dayton, Ohio. I teach scuba diving. Also, I am working on getting my private pilot license, if the weather will ever start cooperating with me! I love retirement!

Chapter Two – The Cold War

C**old War** [kohld] [wawr] – *noun* - A state of political tension and military rivalry between nations that stops short of full-scale war, especially that which existed between the United States and Soviet Union following World War II.

Busting the CORI
Greg Davis

I started flying the B-52G in June of 1989 at Castle AFB, California, and after getting qualified in the aircraft arrived at my duty assignment at Loring AFB, Maine, in November of 1989. At the time I showed up, the 42nd BW was in the process of going through a Conventional Operational Readiness Inspection (CORI), so I was told to go get my family taken care of and come back the following Monday after the inspection.

Now the background to this story goes like this. The Wing Commander at the time carried around a black book in which he kept his naughty and nice list. As I understand it, the aircrew members of the 42nd BW personnel loathed him and were willing to tube the CORI just to ruin his career. To the personnel at Loring at the time, six months of pain from the Combat Evaluation Group (CEVG) would have been a gentle breeze compared to living six more months with that Wing Commander. On the bombing run, the bomb crews would hit the bomb scoring tone at IP inbound and cut the tone right after doing so, yielding bomb scores so bad it was no mistake to headquarters the wing was in trouble. The Wing Commander got the message and called his sugar daddy in personnel who bailed his butt out of being at Loring when the CORI was flown.

With one month to go under the leadership of a new commander, Col Terry Burke, the 42nd BW knuckled down to try to get their stuff together. For the crews of the 69th BS, that meant getting proficient in conventional bombing, naval mine laying, and Harpoon missile tactics in just over a month. By the time the CORI came, the 69th managed to do well in the Harpoon missile and mine laying phases, but the conventional bombing grade was only a "Marginal." We were given six months to get our stuff together.

I had to go through the Mission Qualification Training (MQT) to get certified and, while doing so, also found myself learning about the politics and history of the wing. Every training mission flown by an MQT crew had to go through a major review to prepare for the re-fly of the CORI. All our bombs had to be dropped within 200 feet of the targets and our scores were in front of the SAC Commander within 30 minute of weapons release.

The B-52G had three onboard systems to record the performance of the crew. Besides the 35mm film camera which recorded the bomb-nav system, there was also a video recorder doing the same, along with a B-17 camera in the bomb bay to show the shapes leaving the aircraft. On one training mission, my crew had an additional crew member with a handheld video camera in the IP seat. We had the airplane lined up on the target and were on altitude and on speed at 400 feet Above Ground Level (AGL). The first shape landed at 3 o'clock and 350 feet from the target. It came out of the bomb bay and took a right turn. My reaction to the score was "What?" We came around for a second pass and the second bomb hit at 4 o'clock at 385 feet.

When we got back, Lt Col Dave Capotosti asked us what had happened. I told him we flew over the target. He said that we would review the films the next day. The next morning, we learned the 35mm film had broken, the video recorder failed, and the B-17 camera's film also tore. The only thing that really recorded our performance was the video camera used by the crew member sitting in the IP seat. The handheld camera showed we had gone right over the target with it centered between me and the co-pilot. Both Lt Col Capotosti and Col Burke saw the film which was taken from the IP seat. Col Burke told us we did exactly as we were supposed to do and he would forward the film to Gen Chain if he wanted to see it for himself. It turned out the bomb had a hung fin which caused it to make a right turn as soon as it departed the bomb bay.

The airplane I got for the CORI conventional bombing exercise was a B-52G tail number 6488 with nose art of the Shady Lady. I did not find out the history of this aircraft until after the mission of the re-fly itself. The last sortie it had flown had been six months prior during the original CORI. The airplane had experienced three high-speed aborts and several other maintenance aborts before the scheduled re-fly. My crew got to the airplane and did our preflight, finding we were armed with one Mark-81 Snake Eye inert shape in the bomb bay. We were the number two airplane for the day behind S-01, and our flight was great. We threw the wing's best bomb of the day, hitting only 108 feet from the center of the target. We won't get into the Circular Error Probability (CEP) for the aircraft, but we were bombing very accurately.

When we got back to the maintenance room for our debriefing, we had the forms filled out and the plane was Code 1. The SSgt sat down to take our maintenance debrief and in disbelief said, "You flew 64-Late-Late (the maintenance crew's nickname for 64-eight-eight)?" I answered, "Yes, and it threw a good bomb." Again, she stated in disbelief said, "You flew 64-Late-Late?" Again, I repeated my answer and wondered why she was talking like she did. She explained that the airplane had not flown for six months - no matter how hard the maintainers tried to get it off the ground. I simply said it wanted to fly that day and show its sisters how it was done. I found out an hour later in the O'Club that we had the best bomb of the day ahead of S-01.

The real surprise came the next day when one radar navigator Lyn Scott with the call sign "Magic" dropped an eight-foot bomb. Six months before, Magic was paired with an Aircraft Commander (AC), with the call sign of the Hamster, who could not hit his plate with a fork if it was glued to the table directly in front of him. At first, the Wing Bomb-Nav shop thought it was a Radar-Nav problem so they separated the AC and the Radar-Nav to find out what was going on. Magic, the Radar-Nav, started regularly scoring bombs inside of 50 feet. The AC had been the problem issue. The Hamster was a MAC transplant who was supposed to be on the fast track for promotion. He had to get some BUFF stink on him for career credibility. He had tried to fly through a thunderstorm on an ORI flight after his Radar-Nav had commanded an abort for weather and the Hamster wouldn't do it. The Radar-Nav informed the Hamster the crew was now busted and if he did not abort the route, his actions would take the Wing down on the inspection. The Hamster finally aborted the route and flew home. Everybody knew the Hamster had tubed his crew and the verdict would come out the next morning. The Hamster in panic called his own general sugar daddy and got bailed out by the morning and the crew was miraculously un-busted. His credibility was shot with the crews after that stunt. If memory serves me correctly, the Hamster got to sit out the re-fly because they moved him off a crew and into a supervisory slot.

We wound up passing the CORI with a good grade, and Col Burke told me that if they had wanted us to learn to bomb inside 100 feet regularly, we could have probably figured out how to do so. Our bomb scores for the wing were averaged at about 135 feet with almost all the bombs hitting inside 200 feet.

Alert Scramble - Loring, Maine, 1975

Jay Lacklen

While trapped in the alert facility, the crews kept constant watch on the parking lot located across the street on a small hill 100 yards outside the facility's front door. When a white-topped staff car arrived and parked there, it was our tip-off the Klaxon was about to blow and send us scrambling to our aircraft. One of the colonels would observe the scramble from this optimum vantage point. One day, Col Patterson's staff car pulled into the lot and the word spread rapidly throughout the facility. Sure enough, a few minutes later, the Klaxon blew and away we went, racing to our alert six-man pickup trucks.

For that particular exercise, the aircraft were located on a temporary alert pad used while the regular pad was being resurfaced. The area of the ramp had been surrounded by a chain-link fence, and it would be the first Klaxon alert with aircraft parked there.

An alert engine start was all frantic motion and noise, somewhat like a NASCAR pit stop. Each crew's alert vehicle raced to the pad through the main guarded entrance and pulled in behind one of the wing tips next to their aircraft. The crew scrambled out of the truck and raced to the entry hatch beneath the fuselage while the crew chief ran to pull the wheel chocks. Within seconds of the pilots disappearing up the hatch, the engine start sequence began. Gas cartridges were exploded into the two center engines (#4 and #5) to spin them up to speed as they gushed gray cordite smoke enveloping those engines. As soon as those two engines reached power, the other six engines began to spin simultaneously as bleed air from #4 and #5 surged through the wing ducts to reach them. In quick succession, engines from all four to six bombers on alert began screaming in unison. The senior crew was supposed to taxi first, but if any other aircraft was ready to go and lead had not moved, they could call on command post frequency that they were taxiing. ("Bongo-Three is rolling!") Then the elephant walk commenced to make the 15-minute timing to cross the hold line before taxiing back to the parking spots.

On that day the exercise was not a "mover," just an engine start, but still proved sufficient for drama. First, the co-pilot had an incorrect switch setting for the cartridge start. All eight engine starter switches on his side panel had to be turned off when we had previously left the plane. His imperative first step upon his ass hitting his seat was to flick all eight toggles up to the "start" position. He had not done so when I fired the cartridges. If the two inner engine starter switches were not up, the engines would spool down after the cartridges had discharged and we would have missed our timing and be waiting for a power cart to start the engines. Doing that was a blasphemous error that would have had me chewing carpet in the wing commander's office. The co-pilot caught the error just in time, ramming the switches into the correct positions as we held our breath to see if the engines would continue to accelerate or reverse course and start to catastrophically spin down. They caught, continued to accelerate, and we were saved.

After the engines were all at idle and we had established our timing, Col Patterson's staff car came creeping down the centerline of the pad toward our aircraft. He stopped in front of us and came up on the command post frequency. "Capt Lacklen, you've blown down the fence behind you; what are your intentions?"

I didn't yet know Patterson well enough to know he was mischievously pulling my chain. I wondered if he wanted me to send my crew out to fix the fence.

"I, ah, guess we'll put it back up, sir," I said.

That is when he laughed and said that would not be necessary, and I knew he was playfully toying with me – a rare humor in a command usually devoid of humor."

Harpoons On The BUFF

The Rest of the Story

Mike Loughran

As with everything else, there is a lot more to my story than one ever imagines. This chapter goes way back to the process of how the BUFF was issued a new weapon - the Harpoon missile. This author played a small role in the story and is friends with Tom Goslin, the other officer mentioned, who I am sure doesn't mind getting credit for what he did to make this happen. Later Lt. Gen. Thomas B. Goslin Jr. retired as the Deputy Commander, U.S. Strategic Command, Offutt Air Force Base.

Maj "Goose" Goslin and I were action officers in the Pentagon in an office called the Strategic Offensive Forces Division (XOXFS) which was part of the Directorate for Plans. Our collective chore was to develop force structure options for the leadership to debate about which way to go in the future. In actuality, we did a lot of busy work, but at least our force structure options were projected to be on board in the next 10 or 15 years, not the next day. There were a lot of discussions with our Strategic Air Command (SAC) Headquarters' counterparts who just wanted the Air Staff to fund it all, because "the

CINCSAC wants it!" Hours and hours of time spent with budget types, analysts, contractors and writing papers did not allow much freethinking.

Every once in awhile, however; a real opportunity did pop up. None of us really thought that the war between the British and the Argentines would affect the capability of the B-52 fleet, but it did. The Air Force always had a collateral mission of assisting the Navy, and, in fact they were doing some training at the time in things like sea surveillance and mine laying. Both missions were pretty well matched for the BUFF airframe. One day when we were discussing the Argentinean successes with the Exocet anti-ship missile, Tom Goslin, then a Major, had the idea of resurrecting the B-52 as a Harpoon carrier. It was a project that Lt Col (later General) H. T. Johnston had worked in 1977 when he was an XOXFS action officer.

At the time, we may have been aware of the fact that the Argentinean AF did the whole thing as a "self-help" operation. Later writings in professional journals and after action reports told how the Entendard pilots had never even shot a missile in practice and devised the whole program as a total ad hoc response to the situation.

The reasoning we used was straightforward and can be summed up easily – an airplane is an airplane. The B-52 carries a lot of stuff already and the Navy has Harpoons; so, why not borrow some, shoot 'em at something, and see if it works. It all made perfect sense to a group of Majors in the Air Staff on an otherwise boring day. Since the Goose was the" bomber guy" and I was the tanker guy, when it came to dividing up the workload, he got it! And he made it work.

A sympathetic former B-52 crewmember in the Office of the Secretary of Defense (OSD) staff had control of a large pot of money for research and development, including near-term added enhancements to the existing weapons systems. Being easily convinced it would work, he received some no-name, no-source advocacy papers allowing him to convince his bosses and – surprise! A test program was born.

In fact, some of us were so confident the test would succeed we basically told the Air Force to develop a limited operational capability. The USAF would have three 42d Bombardment Wing airplanes and up to five crews ready to go in harm's way with the Harpoon by the end of September 1983.

Coincidentally, this author then received a dream assignment - except for the location. I was to go to Loring AFB, Maine, as the 69th Bombardment Squadron's operations officer and make the plans happen! The ops officer was responsible for the day-to-day flight training of the aircrews. Tom Goslin's reward was to stay at the Pentagon for a bit longer and fight the really big battle in negotiating the memorandum of understanding with the Navy on "Joint USN/USAF Efforts to Enhance USAF Contribution to Maritime Operations."

The first work ups in activating the Harpoon were typical of any new system's introduction to the field. Ground training classes were developed to explain to the aircrews and ground crews just what a Harpoon was. There was a large dose of help from our friends in the Navy on this effort, because they already had the knowledge of the weapon. It was, however, up to us to develop the Air Force tactics for our use of the system.

We started to work up more mine laying training with sophisticated Navy weapons such as the GATOR and CAPTOR mines. Most of our previous training with mines centered on "dumb" mines, which were basically a 500-pounder with different fuses from the ground bombing version. These larger weapons were truly in the category of smart weapons and required precision in laying the minefield since no one wanted friendly ships to sail into an area of killer mines. That premise made perfect sense to us land attack types, so we worked hard on very precise drops of strings of mines. Precision was something we trained to in every type of weapon delivery.

About that time, the Navy ordnance personnel started showing up at Loring to assemble the weapons and help our blue suit "ammo" guys learn the ropes. Of course we helped the Navy people learn their way around an aircraft a bit bigger than the types they dealt with daily. The initial plans would have the Navy transport the real mines from their storage sites to our base for assembly and uploading.

At first glance, we thought we'd solve this ship attack problem easily. After all, the 69th was selected to become one of the command's premier non-nuclear squadrons performing a wide range of contingency operations. Attacking something on the water would be like bombing in your backyard. Early thought centered on a mutual targeting scheme.

Since B-52s were already flying ocean surveillance missions in pairs, we would use a two-ship hunter-killer arrangement. One aircraft would look at the ocean surface, pick out the bad guy's capital ship, and the wingman would simply kill it with a missile. The plan was for the shooter to drop down low while the high bird would transmit coded messages updating the surface picture, or SURface PICture (SURPIC) in Navy terms – the first in a growing string of USN acronyms we would encounter. Both aircraft would stay outside of SAM range and even though enemy search radars might acquire the high altitude aircraft, an engagement was not likely.

One of the earliest "test" flights of the Harpoon delivery tactic was a rather simple approach to the tactical solution. There was a Navy destroyer coming out of the shipyard at Bath, Maine, after undergoing some sort of overhaul. As I recall, they wanted us try out some ECM gear against a cooperating target that had jamming capability. The Fightin' 69th got the call and added a couple of hours into one of our training missions with the promise that we could attack the ship at the end of the ECM testing period.

The Bay of Maine was like a lake that day and one could easily spot the ship from altitude. Our B-52 crows (Electronic Warfare specialists) and the Navy's crows seemed to enjoy the jamming session, but the pilots and navs waited anxiously to try out a couple of "Harpoon" runs. We had only just basically read the brochures – how fast, how far, how big, etc. since we did not have any missiles yet. The situation was a bit like reading an advertisement for a new car, walking up to a vehicle and jumping in and taking it out to explore the performance envelop.

Our plan was simple - fly low, simulate a shot, and then assume the missile's flight profile and home in on the ship. We knew the ship's position, its heading or Mean Line of Advance (MLA in USN parlance) and, by using straight dead reckoning, figured a likely position where the "missile" would intercept the ship. There were a couple of key points involved, such as when the ship would detect us with our radar turned off, would the known azimuth of the missile's seeker acquire the ship, and an important question, when would the ship's fire control system be able to detect and engage the BUFF?

Of course engagement by the ship was not a concern in an actual shot, because the missile launch point would be well outside the SAM and AAA envelopes. The BUFF's huge radar cross section dwarfed that of the Harpoon, but it would be nice to know all that information. One added bonus was that it was a chance to buzz the destroyer – legally. And I don't care what you say; buzzing a ship, a town, or a group of people is something every pilot looks forward to!

At 500ft (152m) and 300 knots, the ocean can be deceiving. The total lack of features, late in an otherwise gray overcast day, forces you to count on the radar altimeter for height information. We were at the planned flight parameters, well away from the destroyer, and on a heading that would make an intercept easy after we assumed the "Harpoon's" flight profile. I think the prediction was that the ship would not be able to see us until we closed to somewhere around the 20-mile range. That was problematic when the brochure said you could shoot from 60 miles away. And we did just that. After the launch point, the aircraft's speed was increased and its altitude lowered a bit. At a point where the navigator calculated we should be able to see the ship off our nose, there she was. They had turned a bit from the last heading we had for it, but the FLIR sensor picked up the destroyer at about 20 degrees off the nose.

One short radio call let the ship know we had them in a successful intercept. Since we had planned to fly over the "target" ship, I asked for their preference as to which side of the vessel we should "pass in review." Until that radio call was made to the destroyer's Combat Information Center (CIC), they had no indication of our presence and had not detected us with their search radar. So, a missile would have flown for some 40 miles before the target vessel would have a chance to detect it. We made a couple of small heading corrections pulled a hard turn to do a flyby down the starboard side. After a quick circle around the "target", it was climb power, pull the nose up and head for "feet dry." The skipper came up on the frequency and passed on his compliments on the brief airshow.

In the next levels of discussions among the tacticians at the squadron, we generally agreed to a BFO - a blinding flash of the obvious. A cooperating ship on a calm day was going to be a totally different situation than a Soviet Surface Action Group (SAG) capable of inflicting a lot of damage to the attacking aircraft. So, how did we propose to resolve this dilemma?

We decided our Navy friends might be a useful source of information. Either through SAC Headquarters or through direct contact with a Navy P-3 unit, the 69th came up with some tools to aid in solving the attack problem. The vital piece of information would be the data contained in the SURPIC. This became a coded message passed from an aircraft that had "mapped" the surface ships on an area of the ocean. The targeting airplane would fly by an area of interest based on some external cueing from an intelligence source. The grouping of ships that were generally heading the same way in the open ocean areas were probably part of a formation out to do harm to someone. Plotting that group, coupled with a little knowledge of the other guys' tactical formation, would yield enough information to plan an attack.

For example, the MLA, speed, and spread of the group were clues as to the location of the capital ship. In US Ocean transits, SAGs would spread in a relatively predictable manner based on the activity at the time. The SURPIC would try to reconstruct that pattern and number the ships and therefore select a target. A suggested run-in heading would be developed based on the defensive spread of the formation. The attackers' objective would be to avoid the ships, which could engage the aircraft prior to the launch point. Of course, an egress route would also be part of the shooter's planning. Armed with this raw data, then, the approximate Harpoon launch point would determine the routing also.

Well, that was the theory, now to the practice. Two BUFFs would try this approach as a lethal end game to the SAC's Busy Observer sea surveillance missions. There were two types of missions – training and operational; the difference being the Busy Observer II missions went after actual Soviet naval vessels.

From time-to-time, BUFF units were tasked for Busy Observer missions, usually a training sortie with just two airplanes, lots of open ocean, and the chance to find ships, exercise the command and control reporting system and buzz the ships - legally, of course. One BUFF usually mapped, stayed high and watched his wingman drop low to photograph the ships. Then, they would trade places for the purpose of sharing the training.

One interesting sideline to the training sorties was the search for drug runners. On occasion, there would be a specific description of a suspect ship passed along. On every mission, the sighting of vessels that fit a profile would be passed along to the proper authorities for more immediate action.

On a Busy Observer II, our adversaries were out there and we had to find them. A good place to start was the Greenland, Iceland United Kingdom Gap or GIUK. Soviet ships en-route to Cuba usually passed through this gap and this intelligence was passed on to USAF units. However, these sorties did come with a different set of rules of engagement.

Usually, the aircraft carried full loads of chaff and flares as a defensive measure. The on-board ECM gear was tweaked against the specific ship fire control radars as a prudent self-protection measure. The objective was to be fully prepared in the event the Soviet skipper got really agitated and started some sort of firing solution involving the BUFFs. Defensive actions were authorized if engaged by the ship's fire control radar. These exercises built up a modest experience base from which to further refine tactics applicable to the Harpoon mission.

The 69th's immediate task was to achieve a limited combat capability with the Harpoon by late 1983. That was defined as three airplanes capable of carrying the missiles and five qualified aircrews. The aircrew training was basically just what we defined it to be, and it had to be modified as we learned about the weapon system, refined the tactics, and applied the lessons we learned along the way.

For example, the aircraft flight manual had to be changed to include the new equipment, the operational envelope for the missiles, and aircrew procedures. Most of the information was published with a disclaimer noting that it was preliminary and subject to change. Boeing sent two fight test engineers to gather data on aircraft performance with various combinations of missiles loaded on the external pylons.

Generally, dummy missiles were carried on these test flights - mass simulators that duplicated the shape, weight and drag of the actual missiles. Tests consisted of rather boring profiles designed to hit certain data points within the aircraft's performance envelope and note the apparent changes in fuel consumption. The data was then reduced to performance charts for the crews to use in planning missions. One early discovery was an unusual buffeting on some missile control surfaces causing delamination of the surface. The problem was most severe on the missiles at the bottom of both fore and aft stations on the racks. The ultimate fix was to limit the missiles to only shoulder stations – four on each pylon for a total capacity of eight missiles.

Aircrew training consisted of ground classes on system operation, weapons system control, and operational procedures. The mechanical aspects, equipment operation, and "nuts and bolts' was the straightforward part of the training and certification process for the crewmembers. Tactics were a true set of shifting sands because every flight led to a new idea and further refinement of the delivery method of the weapon.

At one point, someone designed what was thought to be a universal targeting method to allow several platforms to search for 33targets, transmit the data to the shooters and insure secure communication between the hunter and the killer. Probably the best example of how that system worked would be to visualize a wagon wheel centered over the target ship, the spokes using letters to identify the potential run-in headings to the launch point. The letter A was at true North and the circle was divided into segments based on the approximate number of degrees of the Harpoon's seeker angle. Target course, or Mean Line of Advance, was an important piece of data. The message passed to the shooter was encoded by various standard methods to insure communication security.

The ideal solution for the hunter was to use a platform optimized to perform that chore, such as an E-3 Airborne Warning and Control System (AWACS). At the time, E-3s were heavily tasked with numerous real world commitments and availability for training was very limited, especially for a B-52 unit working out tactics to attack ships.

The 69th searched for times to practice our growing body of knowledge regarding the Harpoon system. The squadron looked for exercises to stretch their new "sea legs." The squadron thought little of the time involved in flying two BUFFs from Maine to the Caribbean

Basin for the chance to practice just one Harpoon run. In fact, they would willingly be chewed up by F-14s attacking the B-52 simulating a Soviet Naval Aviation Bear trying to attack the carrier battle group.

The squadron learned a number of valuable lessons, including just how far apart the different units of the US Navy were from each other. As part of learning the business, the BUFFs would be under the control of USN E-2 Hawkeye. Navy lingo and jargon was probably as bad as the Air Force's grew to become. One evolution led to the discovery of a new radio language. Navy controllers on this particular asset would tell the BUFF lead to "Push 327.1 - Pogo." Push was a term meaning roughly go to the frequency of 327.1, but the pogo part took some deciphering.

On the first encounter with the term "pogo," our thoughts centered on it being a codeword for some action or another. Perhaps we had missed something in the exercise instructions? Actually, the term was just what it said: "If you don't establish contact on the new frequency I told you to push to, then come back (pogo) to this one." Of course, what else could it mean!

Generally speaking the Navy exercises involving a carrier working up for deployment was a non-productive mission for us. We thought the Navy might like to shoot at some heavies simulating Bears. But the situation was fluid, and the demands of qualifying a carrier air group for a cruise understandably placed our request for a couple of Harpoon runs at the bottom of the navy's priorities - so we went elsewhere.

The first "shot" carried out with an AWACS turned out to be a thing of beauty. We briefed the E-3 crew by phone the day before the sea surveillance and control exercise. We were headed just off the coast of Iceland and near the Greenland area. The objective was to search for a friendly destroyer who was trying to sneak through to engage friendly naval forces. AWACS was the airborne search asset; we were to be the shooter with a single ship using the tactic of looking at the target with a wagon wheel grid superimposed over the ship. The AWACS weapons controller understood the plan perfectly.

After a couple of hours transit time from Maine, the solitary BUFF arrived in the agreed area of operations. There was a bit of surprise when the AWACS answered with a different call sign. Things got a little tenser because it turned out to be a different aircrew too! The original jet broke and these guys took their place on the launch from England hours earlier. Now what?

To our surprise, the AWACS folks did a beautiful job of passing the data from one crew to the other. Our replacement jet not only had it down pat; they found the target too. It turned out that the airplane controlling us was a B-model with an updated radar, making it more effective over water. Information got passed, decoded, and the attack started.

The run-in to the Harpoon launch point was smooth with all crew procedures working well. Missiles were away at the right launch point and course, and the bomber simulated the missile flight and homing. From low altitude, there was no detection by the ship. Right on cue, the gray warship appeared at the 12 o'clock position, and since the ship did not have a helicopter, a quick pass over the stern was appropriate, in the BUFF crew's view.

The NATO planners started to read the literature on the B-52's evolving capability in the sea lines of communication attack role. Requests for the 69th to play in large maritime exercises started arriving in the message traffic via the Commander-in-Chief, Atlantic Fleet (CINCLANT) in Norfolk, Virginia. Most of the training had absolutely too many commuting hours involved. No matter how one decides to travel to the coast of Iceland, Scotland or Norway, it's still a long way!

A trip to Portugal may have set the record for a mission to do a mine-laying run. It was actually a good opportunity to exercise the total system from planning to an actual drop. Training shapes came from the Naval stores and the load crews traveled to Loring to train our blue suit weapons loaders. The mines used had echo locators on board so they could be recovered and recycled for another use. Furthermore, the drop in a little bay on the Portuguese coast was to be the centerpiece of a demonstration to a group of Allied naval officers and dignitaries. A daylight drop of a string of large mines by two BUFFs must have looked impressive to the crowd.

Naturally, the B-52 crews were on time and target for the relatively simple mine run. Mines had to be placed precisely because friendly ships still had to be able to pass through a field. Also, after hostilities, there was a plan to remove the fields for freedom of navigation concerns. This particular run was in shallow water, close to shore and prominent land returns were well within the range of the aircraft radar, making the actual releases a piece of cake for the crews. The ink had hardly dried on the congratulatory messages when the tone changed.

It seemed as though no one could find the mines. The radar film, photos and documentation of the release had to be sent to some USN headquarters for an investigation. Naturally, the first suspects in the case of the missing mines were the aircrew – even though the drop was in broad daylight in from of large crowds, on ships and shore! Well? We fliers had no explanations or excuses. The controversy whimpered on for a week or so and just withered away. We proved the case, the Navy might have found the mines exactly where they were dropped, but records are spotty. In any event, we did it again, many more times.

Ultimately, the marriage of B-52s, Harpoons, and a targeting platform may have reached its peak when the Royal Air Force NIMROD asked to be allowed to take part in an exercise. Number 120 Squadron from Lossimoth, RAF Luchers, got the task. We found a large chunk of ocean and invited those guys over for a stay and a bit of aviation fellowship coupled with tactics development and information exchange. We learned more, I think, than we expected.

The Nimrod MK II, was optimized for the sea surveillance mission – and it was a shooter too. So, they understood the end game better than our own people who had little to no experience in finding, classifying, and engaging ships.

The Nimrod carried Harpoons, mines, and torpedoes for its submarine search and attack mission. The on-board radar was optimized for sea search and was a quantum improvement over the BUFF's according to the guys who operated the B-52 radar. Flying on the Nimrod was a bit like being on the bridge of the Enterprise - the starship, not the carrier. We launched with a two-ship cell of B-52s and the Nimrod.

During air refueling, the RAF pilots just sat out a safe distance and enjoyed the view. It was an easy task to integrate the aircraft into a formation of B-52s en-route to the search area off Nantucket. The eye-opening stuff happened at the merge, as they say, when the fights on. The BUFFs went silent, peeled away from the Nimrod, and the hunt was on.

In short order, the RAF mission director found a warship. His systems could find and categorize ships, almost down to the specific name. Coded radio messages went out to the shooters and we just watched from the relative luxury of the suite of electronics aboard the Nimrod. By then, our tactical considerations had matured along with our thought processes on shooting the missiles.

The 69th's brain trust had done some significant homework in finding out the defensive capabilities and limitations of our adversary's ships. Our objective grew to not just sending in a single shooter, but overwhelming the ship's fire control system with a barrage of missiles. A key point to keep in the forefront is that you did not have to sink the ship to make it ineffective - you could do that with a "soft" kill. By removing is combat capability, knocking out the phased array radars, for example, you effectively neutralized the ship as a combat platform. If you damaged and degraded the ship's combat information center or its ability to command and control other forces, then you had a degree of success in the task.

Additionally, our aircraft equipment had improved significantly in the time since the squadron achieved limited initial operational capability in late fall of 1983. The Offensive Avionics System (OAS) was a big improvement in navigation and attack capabilities. Originally part of the modification to enable the B-52G to carry the Air Launched Cruise Missile (ALCM), the 69th's jets were the first non-ALCM aircraft to receive just the OAS modification. Along with that package came an improved data bus to enable the aircraft to carry some of the soon to be developed smart weapons.

All of these capabilities, in practical terms, led to a much more precise weapons delivery platform. The aircraft could arrive at a point in space and time within seconds of the desired time. In fact, a bomb release could be controlled to the point of the weapon's impact, and detonation, taking place at the exact second the planners wanted. While these improvements were intended for the land attack role, they gave a better tactical use of the aircraft and the Harpoon in the ship attack role.

Thinking had evolved to where we essentially flew a three-ship cell of BUFFs, line abreast with about a mile lateral separation, to a launch point, which was basically a circle centered on the target. Then, each airplane would begin to salvo the missiles at very precise time intervals which were selected to overwhelm the ship's fire control systems. The missiles' extreme low altitude flight profile, coupled with a small radar cross section and high speed, led to a high probability of penetrating the defenses and hitting the ship and causing significant damage to its war fighting abilities. In the largest raid, we thought that out of 36 Harpoons closing very low, very fast and nearly at the same time, some were bound to get through. But, did it work?

The mission with the Nimrod was very successful. Both the RAF crew and our folks were convinced the tactic was sound, the missile would be effective, and the BUFFs were a welcome addition to the problem of going after Soviet Surface Action Groups on the high seas.

In formation with the RAF Nimrod.

Of course, we were all airmen acting like a meeting where everyone was in violent agreement – not always the case in the real world.

The optimists always said, "Of course, it works. We thought of it." But, that approach did not necessarily translate to success unless you tested it yourself against a formidable opponent. An AEGIS cruiser and a battleship would do.

The Iowa, an AEGIS cruiser, and an oilier were crossing the Atlantic and looking for some training opportunities. We got wind of the request through our liaison with the P-3 Orion wing at Brunswick Naval Air Station in Maine. These guys were invaluable to us as they shared a great deal of their extensive overwater experience with the 69th. And, we even flew on one another's airplanes to understand the platforms better and exploit our capabilities in a mutual mission. The Orions were Harpoon capable maritime aircraft that also aided us in targeting and fire control problem solutions. We probably flew as many missions with P-3s as we did with any other asset.

One of the flight commanders in the squadron attended a planning session in Norfolk for this "Sink the Battleship" exercise. The B-52s would own a portion of time when the ships would practice defending against land-based aircraft attack. We were part of a larger force conducting separate attacks against the ships from Navy bases along the coast of New England. It seemed as though all the other jets were Navy attack aviation squadrons flying A-6 and A-7 aircraft accompanied by some EA-6B ECM platforms. Each aircraft type had a block of time with sterile periods to de-conflict the air traffic over the ships. There was even an agreed upon attack corridor through which all the shooters would funnel as they approached the ships. We had a different idea, though.

While the training objectives were to get the ships some air defense practice, we wanted to see if BUFFs could penetrate the defense screen undetected, fly in as close as we needed to be to attack the ships and not be exposed or engaged by them. As part of the agreement, the BUFFs could make two Harpoon attacks; three if there was time. We planned for a third attack, but did not tell anyone. The third set of Harpoon launches points was a piece of information we deliberately did not share with our Navy counterparts. Why? Because, we had planned launches from somewhere they weren't looking for us.

In other words, we approached the ships from their "blindside" by circling around outside of the search radar range and making a run-in from the opposite compass heading from where the ships were looking for us.

While they could detect aircraft coming from the seaside of the ship, they were expecting land-based airplanes to come in on a reasonably straight line from their bases directly to the ship. The long legs on the B-52 did not limit us the same way. In later investigations, it seemed as though the Navy doctrine led them to look in the shortest direction to a land base as the likeliest avenue of attack. It seemed like a very logical solution to a problem, but overlooked the flexibility inherent in long-range, land-based aircraft.

Attack number three was a success – they did not expect it, did not see it; and therefore, could not defend against the missiles with the close-in guns. In the debriefing session, the ship drivers were none too happy with the devious minds of the 69th Bomb Squadron. There was a lot of heated discussion about seeing every Harpoon missile, being able to shoot them down and – besides – the little missile would just bounce off the armor on the USS Iowa anyway. So, there; take that! At that point in the conversation, the Flight Commander, who was also one of the pilots on the raid, simply asked about their expected results against our third attack. Lots of blank stares, hard swallows and agitated Navy guys for that one!

The "Fightin' 69th" worked up to be the premier maritime attack squadron in the Air Force. Squadrons at Andersen AFB, Guam, and Mather AFB, California, followed and built on our early experience. Now, they're all closed.

The body of knowledge is perishable and the time and energies were not wasted. The B-52H now has the sea missions. Equipment from the venerable G-model was salvaged when the airframes went to the boneyard. Some of it sat in boxes somewhere until the Service got around to putting the gear on another BUFF and sending Air Combat Command and now Air Force Global Strike Command aircrews out to sea in a BUFF.

Chrome Dome - Christmas Eve 1966

Pat Branch

Strategic Air Command's (SAC) Operation Chrome Dome kept B-52 bombers on airborne alert from 1961 to 1968. In the event of a nuclear war with the Soviet Union, B-52 airborne alert aircraft had the greatest chance of survival and the greatest assurance of striking their targets deep in the Soviet Union. The aircraft were fully loaded with four nuclear weapons and four Quail decoy missiles in the bomb bay plus two AGM-28 Hound Dog air-to-ground missiles under the wings. Chrome Dome missions were flown over international waters (or ice) and outside the Soviet Union's radar coverage. Because they carried nuclear missiles, it would take an order from the President for the crews to proceed beyond a predetermined (failsafe) point.

Today, reflecting back, the risk of exposing nuclear weapons to accidents seems to me to be the craziest, most outlandish, operation with which I was associated over my 30 year career. And it did happen. I believe in the seven-year history of Chrome Dome SAC lost five aircraft and probably 30 bombs.

From 1965 to 1968, I was assigned to a B-53G crew in the 596 Bomb Squadron, 397 Bomb Wing at Dow AFB in Bangor, Maine. I did not keep records of my flights and after 50+ years there is some fading of the memory, but this is my best recollection. Dow was my first assignment as a B-52 Electronic Warfare Officer (EWO). We went by many names: EWO, EW, Crow, Raven or “Hey You” - all seemed to work equally well.

Maj Charlie Hicks was my crew’s pilot and aircraft commander. Charlie enlisted in the Army Air Corps as WWII was coming to an end. He was discharged but soon recalled for the Korean War and stayed on active duty afterwards. At the time I though the war in Korea was ancient history, but it had been only 14 years since it ended. I suspect that my contemporary, Lt Lowell Kimbel, the crew navigator, felt the same. We were both new in service and on our first assignments. We thought Maj Hicks was old, grumpy, and a very stern master. There was a Wing Commander and a Squadron Commander in the chain-of-command above us but we had no doubt who the real boss was. His name was Maj Charles Hicks. Captain Shelly Beck, our co-pilot, would have been an aircraft commander in any other wing or at any other time. Shelly was experienced and an excellent pilot. In the early Sixties, the 397 BW, B-52 Aircraft Commanders were all very seasoned pilots. Capt Norm Koch, Radar Navigator, had recently upgraded to that position and been assigned to the crew. SSGT LeRoy Johnson rounded out the lineup as our gunner.

We flew six Chrome Dome missions as a crew. Being stationed in Bangor, Maine, we did not fly the standard Chrome Dome track depicted by Wikipedia. Our mission track was optimized to monitor the Ballistic Missile Early Warning Site (BMEWS) at Thule Air Base in Greenland without flying over any other nation's territory. The perceived threat was that the Soviets would likely take out the BMEWS as their first target. The first indication of World War III the United States might witness could be a nuclear mushroom cloud over Thule. This Chrome Dome mission was effectively part of the early warning system. The track took us out over the Atlantic Ocean and up the West coast of Greenland over the Baffin Bay where we would set up a figure eight orbit and keep radar and radio contact with Thule. In 1968 an aircraft accident while flying that same orbit ended the Chrome Dome Operation.

I am writing this story as if all of these events occurred on the same mission. In fact, they all did occur on Chrome Dome missions, but not necessarily on this date or on one mission.

The date was December 24, 1966 (or was it 1967). Oh well, the story is the same. That is the actual date of one of our six Chrome Dome missions. It was Christmas Eve and the planned mission would have us back on base about 0700 hours and home by 0900 hours for Christmas with our families. My sons were young - Mike was three and Tim was two. We had made special arrangements with Santa Claus to arrive at about the same time as my return.

Chrome Dome missions were not much different from standard training sorties, except in addition to the normal routine the crews were issued side arms and a large medal box containing a top secret mission package. Inside the package were the "Go-Code," targets to be struck, and a track from the Chrome Dome orbit to those targets offering the best chance of survival. We had previously passed a mission "Certification" by briefing the Wing Commander on the details of both the Chrome Dome segment of the mission and the combat segment, if so ordered by the President. The order to attack or abort the mission would come in an encoded message over the High Frequency (HF) radio which was monitored by the crew's EWO. SAC and, for that matter, the United States Government treated these missions with the utmost respect. Hopefully, they would be the closest we would ever come to nuclear combat. The side arms we were issued was bit of comical relief. What would you do with a snub nose 38 Smith and Westin when pursued by the Russian Army inside Soviet territory?

This is where the memory begins to fade. Scheduled take off must have been about 0600 hours. I say must have been as that was the approximate time of our planned landing 24 hours later. Unfortunately, on that mission, the landing would not be back at our home base. Somewhere between takeoff and entering our orbit just west of Thule mechanical problems made the pilot elect to shut down engine number two and the electrical generator powered by that engine. No problem there - we still had seven engines and three generators.

On our first three Chrome Dome missions, we flew with the basic six-man crew. Later, after a mid-air accident between a B-52 on a Chrome Dome mission and its accompanying KC-135 tanker over Spain resulting in the loss of both aircraft and at least one bomb, Chrome Dome missions were flown as an augmented crew by adding a third pilot. On B-52G (and I assume B-52H) aircraft the routine on these long missions was to have a qualified pilot in the right seat and either the gunner or the EWO in the left seat. The job of the gunner or EWO was to monitor the radios and ensure the pilot was awake.

We had just settled into this routine and it was my turn to sit in the pilot's seat. I felt like a 14-year-old boy who was being allowed to drive the car. I had gotten a cup of hot coffee for myself and Capt Beck the co-pilot. Major Hicks was asleep lying on the floor behind the pilot's position. You can see what is coming, can't you? Yes I did. I spilled that hot coffee right on Charlie Hicks' bald head, who came off the floor screaming like a stuck pig. It was back to the EW position for me.

My young career was over - there could be no recovery. The previous week, we had been on alert when the klaxon sounded. It had been a Coco Exercise ("Elephant Walk") and the crews had 15 minutes to get to their assigned aircraft, start engines, and taxi to the end of the active runway. Lowell Kinbel, the Nav, had been away from the alert facility for an appointment. I do not remember where he was and it does not make any difference, either then or now. The crews could not leave their aircraft until it was parked and properly back on alert status after the exercise. Major Hicks always insisted on being first. By the time the nav showed up we were the last aircraft to taxi. That meant we would be the last crew back for lunch or to the cribbage game or whatever. There was hell to pay. (Poor Nav.) But this time I had spilled hot coffee on the old man's head.

Later in the mission I was back in the pilot's position monitoring the radios and watching the co-pilot do his work. You can see how highly important this job was. Suddenly a bright red warning light on the overhead right between the two pilots' positions came on accompanied by an audible alarm. It was the Fire Warning light for engine number four. My immediate reaction was to punch Capt Beck on the shoulder. Shelly calmly looked at the warning light and from memory ran the engine fire checklist. Then we had two engines, two generators, and the hydraulics on the left side of the aircraft shut down. No problem, we still had six engines, two generators and the right side hydraulics operating. Capt Beck then told me to go wake up the pilot. After the coffee incident, that was a problem.

After waking the pilot I returned to my normal position. Over the intercom the two pilots and Radar Navigator assessed the situation, what was shut down, what was still operating and if any of that affected the mission. There was no conversation of notifying SAC headquarters or Thule Control. After a due period of silence, I called the pilot and asked him if I should use the HF radio to call SAC Headquarters and tell them what was shut down and that we were continuing the mission. There was a long pause before Major Hicks answered, "Yes, EW. You call them and tell them. Don't ask. Tell them we are continuing the mission." I learned several lessons that day. Don't spill the coffee and do not ask, but tell higher headquarters what you have decided to do.

The remainder of the mission was flown without incident, and our replacement crew had arrived on time. The weather over Thule had been overcast and unbelievable cold - a typical winter's day. As we approached the United States' air space we began to pick up radio calls warning us to be on the lookout for Santa's sleigh, reminding us it was, in fact, Christmas morning. It was then we found out we would not be returning to Dow Air Force Base. All of New England was being hit with a major blizzard. The closest air base with acceptable weather was Charleston AFB, South Carolina. The Wing Command Post would call our families and tell them we were not landing at Dow and that they would get us home as soon as the weather breaks.

We landed at Charleston Air Force Base about 0700 hours. We were met by the Wing Director of Operations who was no happier to see us than we were to see him. It was Christmas morning, manning was at a minimum, and then here comes a B-52 fully loaded with top secret war materials, crypto documents, and a full load of nuclear weapons - to say nothing about two engines, two generators and half the hydraulics shut down. Flying with those malfunctions was one thing, but the aircraft certainly would not be taking off with those systems inoperative.

After securing the classified material in the Charleston Command Post, we were all tired, dirty and hungry. The only place on base open and serving food was the NCO club. In 1966 Officers did not eat in the NCO club, but just that one time they would make an exception. Oh, but flight suits were not allowed in the club - another tradition that has long since died. The club opened a small side room where we would be able to get something to eat. The bar in the NCO club was open and had been open all night and there were several people in there that had been there all night. It did not matter; there would be no beer or alcohol for us by order of Maj Hicks. We felt bad, dirty, and tired, but what was really sad was that on Christmas morning those fellows did not have a better place to be. If the base could turn our plane around and the weather cleared up we were going home.

We did get home later that same day. I have long forgotten what time we arrived back at Dow. Santa Claus arrived at our house some time very early on December 26th. That may have been the first but it would not be the last time that Santa gifts for Mike and Tim arrived on a flexible schedule.

Chapter Three - Emergency

Emergency - [ih-mur-juhn-see] – *noun* - a state, or need for help, created by some unexpected event.

Official USAF Photo

B-52H C'est La Vie

Joe Mathis

We were flying a two-ship cell of BUFFs on just another TDY training mission from RAF Fairford, UK, back in early 1998. We made our way southwest to the Strait of Gibraltar and proceeded east across the Mediterranean Sea with the purpose of making our presence known in the eastern Mediterranean Sea and southeastern region of Europe. On our way to the planned destination in Turkey everything was proceeding as normal, just like several other such missions which proceeded it for a couple of weeks. We were flying the standard planned route and Altitude Reservation (ALTRV) with the outbound air refueling over the Mediterranean, as well as the return air refueling. This mission had a far more significant diplomatic purpose from the earlier ones because on that day we just happened to be carrying a partial load of MK-82 bombs intended for a live drop at a bombing range in Turkey.

Everything was going as planned until we reached the eastern Med and we prepared to enter Turkish airspace. That's where Major Murphy (of Murphy's Law fame) unexpectedly took over as the Mission Commander. The planned live weapon drop had received the blessing of the Turkish authorities and when we took off we had no reason to believe anything would change - but it did. We never received the clearance to enter their airspace. We explored as many options as time and operational abilities allowed, but we eventually had to abort that part of the mission and move on to the return phase.

Of course, we were all pretty bummed out because the live drop got cancelled which added to the downside need to secure and land with live weapons on board. Oh the joy of looking forward to that after an already long day and half a night. Nevertheless, we pressed on and began the communication process with Ops back at Fairford. That's when we discovered that Major Murphy had received a field promotion - straight to General!

Take one part England, one part winter, mix real well, and you get poor visibility. Occasionally you get just the right conditions when the visibility gets so bad your tankers can't take off. Our tankers' launch window came and went with no change in the visibility. They were stuck on the ground, and we were stuck in the air trying to conserve fuel. We still had several hours left to fly back across the Mediterranean Sea, through the Straits of Gibraltar, and on to England. So we were moving on to Plan B, or C, or D. Fairford Ops had earnestly started working on other options long before Plan A had been lost but things continued going South in a hurry. There were no tankers available from England due to the weather and the secondary standby tanker at Aviano AFB, Italy, had already been re-tasked and was en-route to a group of fighters due to a Red X on the primary tanker. The fighter had to have their go-juice before they literally became lawn darts. Unfortunately, the primary tanker at Aviano was not going to be cleared off the Red X anywhere nearly in time to help us, if at all.

Just about that same time word came that the Azores was not what we hoped to hear either. They had a tanker available but getting it loaded with enough fuel and the crew airborne was going to take so long it would require us to fly west to meet them, or at least orbit for a while west of Spain before the A/R. That option made no sense at all because it did not allow us any margin of error at all if something then happened to the tanker or went bad with the A/R itself. We wouldn't be able to get to Fairford and have any fuel for loitering in the area if the visibility remained below minimums to land or divert to any other potential location. We barely had enough fuel to get to Fairford and have a reasonable chance of diverting anywhere else as it was originally planned. That would only be possible if we abandoned the ALTRV and flew direct from our location in the Med, south of Sardinia, to Fairford!

We went through all the scenarios, put our collective heads together as a BUFF crew, and concluded that in order to safely get to Fairford required some drastic steps. Once we got there we needed to be able to delay a short, but reasonable, amount of time waiting for the weather to lift and still have fuel remaining to shoot an approach. If we still could not land we would have to climb out to meet the tanker which was going to launch from the Azores and fly directly to Fairford. Just-in-case Mission Commander Murphy went full bore and we had to go meet the tanker for an emergency A/R, we were going to have to fly direct from our current location to south-central England. In other words, we would have to overfly France in a B-52H, at night, with a bomb bay full of live conventional weapons.

The obligatory radio call was made stating we were departing the ALTRV due to an in-flight emergency and we turned right toward the south of France. A call was also made to Ops with our plan and ETA to Fairford. Ops went to work making arrangements with the tankers to make sure they launched and insured appropriate entities were aware we were headed their way. We got to work picking a direct path from the southern coast of France to the English Channel which allowed us to avoid populated areas in France as much as possible. Being the crewdog geniuses we all were, we knew the presence of a B-52 bomber over France was not exactly going to be well received; and, it wasn't.

As we flew closer to the southern coast of France we made sure we were all ready for any eventuality, so we were strapped in and wearing our brain buckets. We also conducted an impromptu crew briefing and reminded ourselves to maintain perfect radio and intercom ops, as well as to be very careful about what we "might" say. We sure as hell didn't want to let it slip we were carrying weapons over France in a B-52 bomber at night! Soon the conversations with French ATC began and even though we all knew we were not exactly going to be welcomed with open arms, the immediate tone and vociferousness of the first French controller was still somewhat surprising; and things just went downhill from there!

The radio communications turned into a very entertaining and ongoing exchange with numerous ATC sectors as we made our way over France. Of course, the French intercepted us with fighters so we had our obligatory communications with them throughout the crossing as well. Each time we were handed off to the next sector of ATC, the entire communication circus had to be repeated for the most part. Each time there were several individuals and supervisors who had to have the situation explained to them. This was apparently needed because none of them could believe a B-52 bomber was actually flying over their nation in the middle of the night! Go figure! The tone of disbelief in their voices certainly painted a picture of what was going on. Even after all these years, it still makes me chuckle to imagine the scene at each ATC location when they got the news of what was going on and the subsequent chaos in each as we made our way over France! The radio chatter was hilarious; I just wish we could have seen the looks on their faces!

Eventually the fun over France came to its conclusion and we reached the English Channel. The constant radio fun with ATC and the French fighters had added to the work load as we had continued to get weather updates for Fairford as well as numerous other locations. We continued working on planning for an emergency A/R, explored options for diverts, calculated numerous BINGO fuels based on circumstances and destination, and loaded numerous potential location waypoints in our navigation systems. We also prepared for potential radio comms at different locations, and other potential distractions. We were going to be operating on the edge, and potentially on several edges consecutively with no room for error, while jumping through hoops with no time to spare. We had been very busy preparing for every eventuality the circumstances were giving us, and we were sure hoping to catch a break in the weather at Fairford and avoid the chain of events that could come if we did not get the visibility necessary to land.

Everybody that was somebody at Fairford was actively monitoring the weather and visibility, and all the airfield assets were ready to roll as we descended into the pattern there. The situation allowed us a straight shot to the field with full cooperation from ATC. That was fortunate because we arrived with just enough fuel to shoot two approaches before we would have to bug out. Thankfully, the visibility started to improve slightly as we arrived and on the second approach it was "just good enough" for the pilots to land the plane. Post-flight activities were completed and our mission debrief conducted, with some added emphasis on the return leg of course.

All things considered, we were a tired, but relieved, B-52 crew which had done something that falls into the category of "rarest of rare" events. We flew a B-52 full of bombs coast-to-coast over France in the middle of the night.

C'est la vie!

Was It The Triangle?

Russell Duffner

The Bermuda Triangle is a region in the western part of the North Atlantic Ocean roughly bounded by Miami, Bermuda, and Puerto Rico, in which ships, planes, and people have been reported to mysteriously vanish.

Back in the Sixties, I was flying as co-pilot with Maj. Ray Tissot on a normal B-52G training mission out of Ramey AFB, Puerto Rico, and encountered some strange navigational phenomenon. Shortly after takeoff, as we climbed out, odd things started happening to our flight instruments. We immediately called San Juan Center, got a block altitude and declared an emergency. We were totally confused when the number one and two UHF radios were working normally for a while and then they wouldn't. The same thing was happening with our TACAN, VOR, and multiple other pieces of equipment. The EW and Gunner were told to turn off all their equipment so we could try to isolate the problem.

At that point, we had a visual of Hispaniola so we kept flying a visual orbit off of it. We contacted the Ramey command post as soon as we could make contact with them. We had to turn both UHF radios to the same frequency since they came and went on and off as we orbited. After numerous discussions and attempts to solve the problems, the engineers at Boeing were finally consulted. Suddenly we noticed the whiskey compass was also not reading correctly. (Bermuda Triangle?) On a heading that should have been about 330 degrees, it read 090 degrees. All of the folks on the ground thought we had must have been drinking the whiskey or were smoking wacky weed!

As the flight progressed, it was decided that since we weren't sure what was and what was not working properly and we should try to get the gear and flaps down before something happened to them. We were happy to find both came down successfully. By then it was approaching nightfall and we (the crew) decided it would be nice to try a landing before dark since we were unsure about what lighting we had. When we informed the staff on the ground of our intentions, the Command Post disagreed with our decision, because we were about 15,00-20,000 pounds over the max gross landing weight. Ray, who was extremely mild mannered, politely informed them to clear the runway as he was coming in and giving himself enough time for a missed approach and a second daylight attempt if necessary.

The approach was perfect and the landing outstanding (unlike most of mine). We parked on the hammerhead, shutdown the engines and unloaded, and as we headed to the crew bus Ray turned as white as a ghost. No one would believe us during our debriefing. Several days later one of the maintenance super sergeants told us they had found the problem and everything jived with his explanation, including the whiskey compass oddity. It seemed a 6-10" piece of copper wire had fallen against the aircraft's BUSS panel causing the wires to melt and fuse; so as we turned or the aircraft pitch changed the wires would sway and come in contact or lose contact with the BUSS. As for the whiskey compass, the electrical problem was creating a magnetic anomaly which was causing it to read in error.

It was just another day at Ramey, and not a Bermuda Triangle entrapment after all!

Official USAF Photo

Near-Death Takeoff
Jay Lacklen

Around the time my two pilot training IPs died in aircraft crashes, I nearly did too - twice.

The first episode took place on a routine takeoff from Loring AFB, Maine, and proved to be a near-replay of the 1969 crash mentioned earlier. As we rolled down the runway, we barely made our acceleration check - the speed that must be reached by a certain elapsed time or the takeoff must be aborted. This point gives the aircraft room to stop in the remaining runway.

This alarmed me because we usually easily beat the time. After eating up 8,000 feet of the 11,000-foot runway, we still remained 30 knots below rotation speed and were barely accelerating. Something was terribly wrong.

I had pulled several of the throttles back slightly to match the prescribed takeoff power setting, but then I shoved all of them to the firewall. At least they'd know we gave it everything we had if we didn't make it.

I looked down the remaining runway and saw the approaching pine forest off the end, where it seemed we were about to meet a fiery and apocalyptic end. At that point, my mind began going haywire. Viewing the remaining 3,000 feet of runway, I thought, in rising panic,

I could easily stop my car in that distance and perhaps I should try to abort on the runway. Fortunately, I also recalled an imperative statement John, my Castle IP, had impressed upon me: "When in doubt, continue the takeoff." John's advice saved us, because if I had as much as touched the brakes, we'd have died horribly in that pine forest.

As the end of the runway approached, I experienced my life flashing before my eyes — in an odd form. I pictured my family members and Boonie Bill's dogs for some crazy reason and marveled, almost serenely, that I now knew where I would die—right off the end of this runway. At the last possible second, I pulled the yoke back hard into my stomach, not knowing if the plane had enough airspeed to respond. It did! We blew all the dust off the overrun and limped into the air barely above the treetops. I briefly considered ordering the crew to bail out since I still didn't know what was wrong or if we could gain airspeed and stay airborne. It was probably the same dilemma that confronted the Aircraft Commander in the 1969 crash. But I realized the navigators, with downward ejection seats, did not have enough altitude for a successful escape. No, we would all make it or not make it, unless we approached a stall. If that happened I'd have to at least try to save the four of us upstairs and order the bailout.

When we turned onto the 17-mile arc as per the departure procedure, we were only 2,500 feet above the terrain. We should have been at 10,000. The co-pilot's eyes were wide as saucers when I dared to look at him. We had both thought we were dead. I aborted the mission, burned down my fuel to the maximum for landing, and put it on the deck with a demand for maintenance to meet us at the parking spot.

Upon later inspection, it turned out maintenance had mistakenly trimmed the four outboard engines to produce about 5% less than their full power. We might have caught that, but we normally concentrated on the two center-engine instruments that showed proper settings. The reduced outboard engine thrust was just sufficient to make our acceleration check but not enough to make our rotation speed before the end of the runway. It was also fortunate the engines provided just enough power to save our skins.

That night, I visited the pizza restaurant located not quite a mile off the end of the runway and just off the extended centerline. I sat down at the bar in my flight suit, but before I could speak, the owner exclaimed loudly to me that some idiot pilot almost tore the roof off the place earlier that morning.

I told him, “Yes, I know; and I knew exactly who that idiot was.”

Excerpt from “Flying the Line, An Air Force Pilot’s Journey,” book one. Book series web site: www.Saigon-tea.com.

The Reason For Our Existence

Roland R. LaFrance, Sr.

Former SSGT,

Fire Protection Specialist (57150) USAF

The B-52 Stratofortress is a very large, complex bomber which, until the late 1990s, carried a crew of six. To the Air Force, and especially all of the aircrew, ground crew, and support personnel, the bomber was an important piece of the Air Force inventory. However, each crew member was more important than the aircraft, because the bomber could not fly without them. On most Strategic Air Command bases the bombers and tankers and other support aircraft were the reason for my existence. I was a USAF/SAC Crash-Fire-Rescue firefighter.

The following story is real. It happened in the 1973. I was there.

The call came in via the Crash Net, from the base control tower. It reported a B-52 was down in the area beyond the overrun area south of the parallel runways at Andersen AFB, Guam. Already clouds, columns of viscous, oily, black smoke billowed up into the bright sky from the burning JP-4 jet fuel surrounding the bomber. The base tower passed along the number of "Souls On-Board" (S.O.B.s) to the Crash Control Room operator at the Crash Station, who broadcasted it over both the station intercom, and then the radio.

It went something like this:

"TEN-TEN-TEN (the radio code for an emergency), All responding Crash and Structural Fire equipment; we have a B-52 down beyond the south end of the runways. Six Souls On-Board ..."

Crews quickly jumped onto their assigned apparatus, including several large P-2 crash vehicles, called rigs. Firefighters responding from their assigned stations or patrol/stand-by positions saw the smoke and angry red-orange flames dancing within the smoke, and headed straight into danger. It was the most extreme reason for their existence.

The driver/operators of each of the 33 ½ ton P-2s steered them to approach the wall of fire and smoke, to best position themselves towards the nose of the aircraft, almost totally obscured by the smoke and flames. The roar of each of the P-2s' twin engines combined with the roar of the flames. For the B-52 crew, time was of the essence. For us, seconds and minutes truly counted. As the P-2s approached the wall of fire and smoke, switches and levers were thrown and suddenly the roof and bumper turrets of each big rig started discharging foam (AFFF = Aqueous Film-Forming Foam). Each rig could discharge up to 1,400 gallons of foam per minute to knock down flames, and smother the jet fuel.

When a large, modern aircraft crashes on take-off or landing the fuel tanks usually rupture, spilling thousands of gallons (tens of thousands of pounds of jet fuel.) Any hot area of the aircraft and its engines can ignite the spilled fuel.

The crews of the P-2s used the turrets to knock down the flames to create a rescue path to the port side of the aircraft. The P-2s approached the aircraft, one on the nose, two to the right side (on approach) and at least one to the left. The P-2 closest to the port side blasted a path through the flames to be positioned for rescue of the air crew in the forward compartment. The crew chiefs and two firefighters dismounted each P-2, supplemented by ramp patrol crews, and pulled the foam hand line hose from the front of each rig, continuing to extinguish fire or protect the rescue path.

The three-man Rescue Crew, having parked their P-10 rescue vehicle off to the side, donned their crash hoods and gloves, grabbed their tools, and headed to the P-2 positioned next to the port side of the bomber. Access to the crew compartment is gained by going over the top, from the P-2's roof, and down into the bomber through a top hatch.

As hand lines continued to put out the fire, and turrets continued to discharge foam on stubborn or more distant flames, hand lines were also used to protect the P-2 in close for the rescue, and maintain the rescue path, protecting the rescue teams.

The B-52F is a huge aircraft, and can carry very large quantities of fuel. The firefighting team knew their jobs, and the rescue team had to know the aircraft well - well enough to work in smoky conditions inside the aircraft, shutting down systems, and especially making each ejection seat "safe" so an unwanted ejection didn't cause injuries or deaths. Safety pins were placed in each seat, or the seats' ejection system initiator hoses were cut.

With the P-2 in the "rescue" position, the rescue crews climbed the ladder at the rear of the rig, to get to the roof. A rescue ladder is stored there to be used to get to the top of the fuselage of the B-52, next to the Aircraft Commander's ejection hatch. One of the tools carried by members of the rescue team is a "de-arming tool." It looks like a pair of bolt cutters, with a large, cutting hook at the end, and is used to carefully cut the thick ejection system hoses attached to the hatch, and if needed to cut the hoses on the ejection seats.

Using the external hatch release handle, the hatch was opened slightly, and the hoses to the hatch were cut. The rescue team moved down inside the cockpit, where they first removed the Aircraft Commander, then shut down systems and then found and safely removed every other crew member. Each rescue firefighter knew in which compartment, and which crew member he must get to. The rescue ladder used to reach the top of the bomber was flipped over, where a rescue slide is permanently attached to the ladder, and is used to lower the crew members to the roof of the P-2. There, other firefighters safely lowered them to the ground, by the Stokes rescue basket stretcher, if necessary.

Inside the dark and smoky fuselage, rescuers worked to free the crew. In their aluminized-asbestos bunker gear pockets were ejection seat safety pins, rescue straps, as well as other small tools. If any crew members were trapped by broken or bent aircraft parts, more tools would be brought in from the P-10 rescue vehicle to use in an all-out effort to save the lives of the B-52 crew.

Each crew member was unstrapped from their "saftied" seat, either by the rescuer using the quick-release, or using a rescue knife, if necessary. Each then had a rescue strap put around his chest, under their armpits, and was then lifted out of their seat and brought to the open egress hatch for removal. Rescuers worked as quickly as possible in the smoky darkness of the crew compartments, working to prevent further injuries. Working down in the "Black Hole", where the Nav and Radar Nav are seated, was the most difficult, and the danger from the downward ejection seats was real. They must be carefully saftied before the crew members are removed.

Early in the rescue for this B-52F it had to be determined where the tail gunner was located. If the gunner was still in his turret at the aft end of the bomber, another rescue team must go there. Using the built-in emergency egress features, they could blow off the turret guns, or enter the gunner's compartment through the entry/egress hatch to rescue the gunner.

Outside, the battle against the flames continued until the battle was won. Fire department tankers and pumpers were brought up to replenish water and foam supplies in the P-2s, and five gallon foam cans from a large trailer were taken to the roof of each crash rig to refill foam tanks with AFFF. Bomb Wing staff were on-scene, watching the rescue and directing efforts, and ready to do what was needed for the fire crews to accomplish their mission. The aircraft itself may be reduced to a burnt-out shell, but it is the lives of the crew that is uppermost on everyone's mind.

The firefighting and rescue operations you have read about may actually have taken less time than it took for you to read this. Rescues must be accomplished quickly, in minutes, if possible.

This incident was very real, but it was only a training exercise. The flames were real, the smoke, the danger, even the B-52F was real. The dedication and the performance of the firefighters and rescue crews were real. Even the extreme danger was real. The only things "simulated" were the crew members inside the aircraft. The crew members rescued from this B-52F, nick-named "The Grey Ghost," were only weighted cloth dummies.

In this exercise, every firefighter and every crash crew got at least two turns fighting the flames and aiding in the rescues. During one of our training "pit" fires surrounding "The Grey Ghost" I was one of the secondary rescue team members, working in the cockpit and atop the bomber.

During this period, every Fire Department on every SAC base held training fires on a quarterly basis. Besides these training fires, there were daily, weekly, and monthly training sessions on different aspects of our jobs. We often did rescue drills inside in-service B-52s, and each other aircraft model assigned to our base. We also had to have a working knowledge of the other aircraft that might visit our base in case of accidents. Sometimes actual crew members were sitting in the ejection seats. We looked forward to performing our best in these drills, ever interested in learning more, and becoming even more proficient in our life-saving skills.

I never had to work the crash of a B-52, but I was certainly prepared. My team was well-trained, with a devotion to duty that leads men to walk into fire to save another. I did respond to a B-52D that was on fire, at Andersen AFB, in 1973. (See page 102 of *We Were Crewdogs VI*, First on the Scene)

In 1975, my training and dedication as an Air Force Firefighter led me to be named Crew Chief of our secondary rescue crew, and anytime I was assigned to work as a crew member on Loring's Rescue-2 was an honor and a privilege. I worked with some of the best, and best-trained Air Force firefighters and rescue men.

We watched over the B-52s, KC-135s, and the crews that flew and maintained them, and all the people who supported them. We did so whether it was an aircraft emergency or a fire in a housing unit. Protecting everyone on base was … The Reason for Our Existence!

My First Solo Flight

George H. Schryer

I had my first airplane ride when I was about eight years old. It was in an old cloth winged Piper cub owned by a returning WWII pilot starting his own local flying service. I got the bug as a child and years and years later I was finally going to get to fly as a real job. The journey had many twists and turns but at last the time had arrived. I had finished Gunner's school at Castle, spent hours on the simulator, flew several missions with an instructor, and been certified as qualified to be on my own by the head Stan/Eval Gunner. My solo flight in a B-52 was finally here.

I had been looking forward to the day for a long time. I didn't get much sleep the night before; I kept going over and over what I was supposed to do and when I was supposed to do it. I guess I must have flown that flight a dozen times in my head before I got up, got ready, and drove to the squadron building to catch the crew bus. We were scheduled for a 0805 takeoff. I stopped at the inflight kitchen, picked up the crew lunches, and loaded them and my gear aboard the bus, then went inside to find the crew. They were all old hands and I knew they would be keeping an eye on me throughout that first flight. I had nothing special scheduled so it was going to be an easy one for me but I was still pretty nervous about trying not to make any big mistakes. That was going to be my crew for the near future and I wanted them to accept me as a competent crew member who could be counted on to do my job in a professional manner. We were scheduled to go on alert the following Thursday and I still had to be certified in that area. I had been hitting those "books" too and I felt when the time came I would be able to give my little specialty briefing without any problems.

The crew finished a few little squadron details and we boarded the bus headed to Base Ops to get our final weather briefing and file our flight plan. Then it was back on the bus and out to the flight line to where our bird was parked. The Crew Chief had the aircraft's 781 form ready for us with no aircraft problems noted, so we loaded our gear and did our walk around before we preflighted our ejection seats and waited for engine start time.

At engine start, everything looked good so we taxied out to the hold line and were given the okay to take the active runway. Throttles up and we began to pick up speed. I can't explain the feeling I had at that moment - finally I was going to begin doing what I had spent all that time training for. Then it happened.

Just as the wheels broke ground it seemed like I went blind. I couldn't see anything - my scope, the EW setting next to me, my hand in front of my face, nothing. No one else in the cockpit could see anything either, which is not a good thing when you are that close to the ground, and the end of the runway is approaching. I heard the pilot tell the co-pilot he had the aircraft and would continue the takeoff profile and instructed the co-pilot to open his side window. The A/C also opened his side window and declared an emergency, stating he had lost visual flight and was continuing on the current heading and climbing while assessing the emergency.

He didn't seem too excited (which settled me down a little) but what the Hell had happened? In all the "war stories" I heard prior to that flight, I had never heard one where everything turned black and no one could see. Pretty soon visibility became a little clearer in the cockpit and the A/C asked the EW to open the sextant hatch to see if it would help suck out some of the smoke, or whatever it was, inside the cockpit. Soon things began to settle down and the pilot and co-pilot started talking to maintenance about the problem. It was decided we had experienced a catastrophic failure of the air condition system's charcoal filter. Somehow it had apparently disintegrated and filled the cabin with carbon dust.

Fortunately we were all on oxygen for takeoff so none of us inhaled any of the dust into our lungs. Still, it was going to be a long flight around and around the flag pole before we could burn enough fuel to get down to a safe landing weight. It was common for B-52 aircraft to take off with much more fuel aboard than they could land with. Finally after endless hours of gear-down, flaps-down flying, the guys up front put us back on the ground. As we taxied back to our parking space we saw a lot of people were waiting to meet us, from the Wing Commander, Squadron Commander, and Maintenance Commander down to the guys driving the hospital ambulance. We looked at each other after getting off the plane and we all started laughing. Each of us had, what appeared to be, a black mask around our eyes from the carbon dust, but where our masks had been, our skin was still white. We sort of looked like a bunch of raccoons.

After the A/C talked to all the commanders we loaded onto a bus and were taken to the base hospital for a checkup. Because we had been on oxygen the exam was short and sweet and soon we were back on the bus headed to the squadron building. After our shortened debrief I got in my car and went home. When I walked into the house I thought my wife was going to have a heart attack. I still looked like a raccoon. She was already not fond of my flying so this sight only strengthened her fear that something bad was going to happen to me someday. I assured her, with all the confidence of a true SAC warrior, that this was just one of those things and it was no big deal.

So my first solo flight consisted of a four-hour trip around and around the pattern. Big Deal! It just proves the old adage that flying is hours and hours of sheer boredom followed by moments of stark terror.

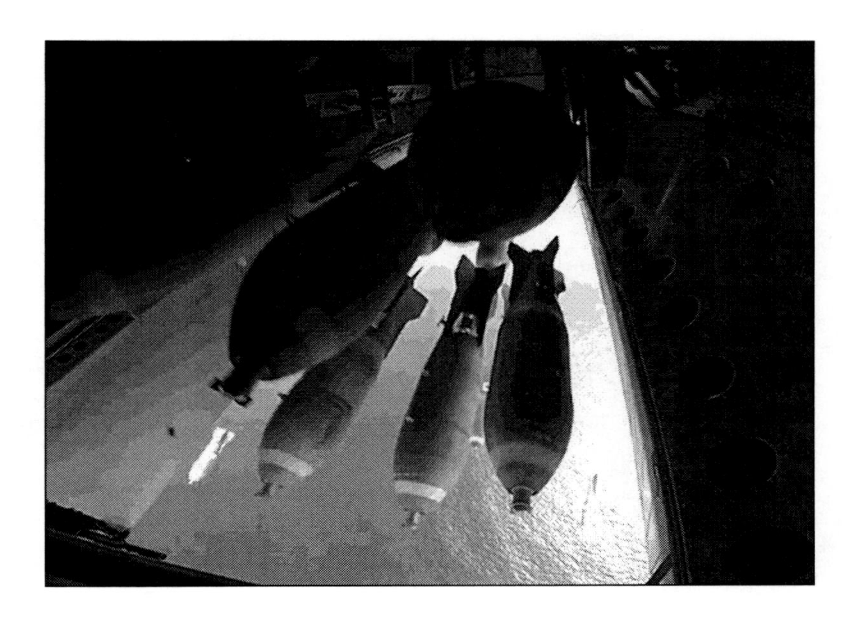

Landing With a (Major) Hanger

Peter Bellone

I was a gunner assigned to a B-52D crew out of Westover AFB, Massachusetts, in 1970 and sent TDY to Andersen AFB, Guam, in early 1971 as part of Operation Arc Light. I don't remember the particular crew number or tail number for the flight but it was a morning flight out of Andersen with normal refueling over the Philippines and then a direct route to the Vietnam ADZ. We were a standard cell of three aircraft and crossed the ADZ flying northward to our assigned target. We hit our primary target and dropped all our bombs as advertised; however, we later found out some ordinance did not shackle and release properly. As usual, the navigator went back toward the bomb bay to inspect for any bombs not released. What he found was very interesting - one of the 750 pound bombs must have released late because as it fell it was grabbed by the bomb bay doors as they closed. So what we had was a 750-pound bomb sticking out of the aircraft, nose down, and being held in place by the closed bomb bay doors crumpled around the bomb. For those who are curious as to how I know this, that will be explained later in this story!

The navigator called and informed the pilot the rest of the crew of the problem, which was a cause for alarm among us. The pilot requested the EWO call this problem to Giant Step which was the radio call sign for the Offutt AFB command post. After a brief wait, the reply was to ensure that the bomb was secured somehow and to report back to them when it was accomplished. The Radar/Nav went back into the bomb bay and did his best to secure the bomb to the aircraft, but he was skeptical of his ability to do so successfully. Our EWO contacted Giant Step with that information and also with a request to go out over the South China Sea and jettison the bomb after clearing the area for shipping.

Giant Step had us vectored over to the coast of Thailand, but further south and out to sea, but we could still see the coastland really well from our attitude. We were there for around an hour or so, and the Navigator cleared the sea area in anticipation of dropping the rogue bomb there; however, that was not what happened. Giant Step said that we could not be sure that we could clear all shipping and other boat traffic in that area, and they were also worried about the Russian trawlers that monitored our actions in the South China Sea as well as off the coast of Guam and Vietnam. Therefore we were directed to make sure the bomb was secure as best we could and to land at U-Tapao Royal Thai Air force Base. That order sent the Radar/Nav back into the bomb bay and according to him; he secured the bomb with the strapping wire used to secure the bombs arming system until release.

We then proceeded to land at U-Tapao. Now what makes this story more interesting, and scary for all of us on board, was that my captain pilot could not land the B-52 smoothly to save his life! In fact, we thought we were going to lose him on a training mission at Westover due to his inability to land the aircraft properly! He usually bounced the aircraft down the runway until the airspeed bled out and yards of runway were used up. Granted, he usually got us on the ground in three or four bounces, and we always could walk away from the aircraft having cheated death once more. But that time he had a reason to land the aircraft in a manner like he never had done before, as all our lives depended on it.

According to the Emergency Checklist, I had the option of bailing out if I so desired, but to me that was not the way to go. I reasoned that if the bomb broke lose upon landing the resulting explosion would take us all out and we would go as a crew. Youth has its bravado!

So there we were, on final approach, and I could hear the pilots talking to each other. Everybody else was quiet as a mouse, which was normal on approaches, but that was no normal approach. To bring home the intensity of how bad my pilot's landings were, I usually placed my seat backward, and put my feet up on the front console to brace for the landing bounces. I found I didn't bounce so hard or as violently when I did so. One needs to understand that in a B-52 when the nose when down, the tail went up, and vice versa. So if the nose came up and down in a violent action such as landing, the tail was at the end of the whip and the gunner was bounced around like a BB in a shaken tin can. For those not in the know, the B-52 has four trucks and not a tricycle landing gear like many conventional aircraft. Therefore to land a B-52 it is necessary to align the aircraft with the runway, as airspeed is bled off, and then slam the beast onto the ground, and hopefully keep it there! Then the drag parachute in the tail is released to help reduce the landing roll distance. With wings as large as the B-52 has, the aircraft wants to continue to fly, so it is a chore to keep it on the ground. Since my pilot always had such a difficult time landing the beast, we usually logged at least three landings each time we touched on the runway!

Finally we were on final approach to U-Tapao airbase with a live 750 pound bomb sticking out of our fuselage and trying not to slam the aircraft into the ground without the bomb going off and sending us to our maker! The tension in the aircraft could be felt even as far back as the gunner's compartment. With the flaps down, air speed and angle of attack being called off by the co-pilot, we all silently said our prayers and said goodbye to our loved ones, as we got closer and closer to the runway. Finally, the moment arrived and instead of being slammed into the ground, this guy who normally couldn't find the runway with both hands, does an airline touchdown. It was a greaser! In fact it was so smooth you could hear the slight screech of the tires, and then the front trucks touching down as well. It was a textbook landing and we all did a sigh of relief when the bomb didn't fall from the aircraft and there was no explosion.

We were vectored to stop as soon as we could while all the emergency vehicles came rushing out to greet us as soon as we turned off the runway and came to a halt. For the first time as a gunner riding in the rear gunner compartment, the expanding stairs were immediately set up and the compartment was cleared to be opened. That happened quickly. At first I thought I was going to have to use the emergency escape rope, but I didn't have to. The ground guys were on the ball and we were out of that tin can in a hurry.

I wrote earlier I would explain the positioning of the bomb and here it is! As we walked around the aircraft toward the crew buses, I could see the bomb hanging out of the bomb bay doors. The length of the bomb nosed downward toward the ground with about five or six feet to spare. However, as I looked at the bomb just hanging there, looking all mean, but defenseless, I realized if the pilot had made his usually bouncing and slamming on to the ground, we would all be statistics and not writing this event as an afterthought!

It has been 44 years since this incident, and shortly thereafter, I left the gunnery field to become an Airfield Operations Manager. I got to see my former pilot again just before he left for Arc Light one more time; however, just before he returned, I was shipped out to Cam Rahn Bay, RVN later that year. I never got chance to thank him for his perfect landing, so a belated thanks goes to the pilot of that aircraft, and a well done to all who rode with her to the ground.

Chapter Four–Southeast Asia

Southeast Asia [south-eest] [ey-zhuh] - *noun* - The countries and land area of Brunei, Burma, Cambodia, Indonesia, Laos, Malaysia, the Philippines, Singapore, Thailand, and Vietnam.

Into the Lion's Den...Again!

Falcon Code 109

Scott Freeman

The flight on the 21st of December 1972 started off much like my crew's first two missions to Route Pack 6, deep into the heart of North Vietnam. When I say it started like our first two missions, it's only because my alarm went off at 1500 or 1600 in the afternoon as it had doing for our first two Linebacker II (LB II) missions. Why that early you might ask? It seems our crew was on a rotation schedule of flying late at night and we generally landed from those missions around 0630 in the morning. We had been so wired and overjoyed just to have survived our missions, we did not hit the sack right away. Actually we did a fair amount of rather serious drinking of just about anything we could get our hands on. We'd sit for several hours talking about the mission, what we did and did not see - since some of us had no windows in the B-52 D-model. Of course, we also discussed what we thought was going to happen on our next mission and how long Operation Linebacker II would last. Our first mission's target, on 18 December, was Radio Hanoi and we hit the Thai Nguyen Power Plant on 19 December. We did not fly on the 20th but were up-to-bat again on the 21st for what turned out to be a very scary mission - one of the worst of the five Linebacker II missions I flew.

I think it is appropriate to provide a bit of background information about our crew. We were crew MAR S-23 out of March AFB, California. We had received our Select Crew (S) rating prior to our current tour and were very proud of it. Our Wing Commander at March was Col Glenn Sullivan, who later became the 307th SW Provisional Wing Commander at U-Tapao as BGen Sullivan. He was an incredible officer, leader and friend, especially to his March AFB crews. On more than one occasion we were invited to his trailer for drinks and sat around just shooting the bull. He was a very special man to all of us and one who went to bat for us to get asinine SAC tactics changed for the 4th night of Linebacker II and beyond. That action probably cost him a second star.

Our Aircraft Commander was a very experienced and exceptional pilot and had done a previous Vietnam tour in C-123 Ranch Hands dropping Agent Orange. (No, he did not smell like an orange.) I would fly into Hell with him. (Wait, I did!) Our co-pilot was rather new to the crew and had joined us in the fall of '72. We had flown a few previous missions to North Vietnam where we were shot at with SAMs, so he had seen them up close and in person. Our Radar Navigator was a Major with many Arc Light tours under his belt. Our Nav was on his third tour, as was I, and was very experienced as well. Our Gunner was a MSgt with a number of Arc Light tours under his belt and knew the BUFF inside and out.

On the day in question, my trailer-mate and I headed over to the O' Club for dinner around 6pm to see if we could choke down some food. The problem was we did not have much of an appetite. One of our other buddies from Westover AFB joined us since he also had the same bus pickup time as we did. Little did we know at the time he would be flying in Scarlet Cell. (More on that shortly.) Why no appetite? I guess I would call it a case of "severe apprehension" about what lay ahead of us for the evening's mission.

With my crew, after our first mission to Hanoi.

I have often been asked, "Were you scared or frightened?" I have thought about that many times and my knee-jerk answer is "Yes." I can tell you one thing for sure. I was never scared during the missions, as I was far too busy to even think about it. I probably would have been terrified if my crew position had a window and I saw the SAMs coming at me. In reality, I focused so hard during the missions making sure my jammers were fine-tuned just the way I wanted them and was constantly scanning and listening to make sure I missed nothing so I could make the threat calls just when I should.

You can say one thing about SAC - they trained us well and, as a crew, that training clearly saved our lives numerous times. Except for the co-pilot, our crew had flown probably 150 missions together so we were a senior crew by the start of LB II. We knew what each crewmember was going to do and could always count on them to do it. They were my "Band-of-Brothers."

I recently read Robin Olds' memoirs, "*Fighter Pilot*" which I highly recommend. Truly he was one of the great Air Force aviators; an officer, gentleman, and leader. I met him once in August of 1973 in Las Vegas at the 1st Red River Rats Reunion which our newly returned POWs also attended.

Indulge me a slight digression here. I purchased a River Rat beer mug and was about to get it filled when someone slapped my back. It was BGen Olds. He had seen my B-52 "200-Mission" patch and said "A BUFF guy. You guys were just flat crazy flying like you did over Hanoi. At least I could maneuver in a fighter." I was so flustered I think and all I blurted out was something about getting our POWs home. He asked what I was drinking and I replied beer. He told the bartender to fill my mug with gin! I thought, "WOW! I have to drink this no matter what," and we talked about LB II for about 20 minutes. I was truly in awe of this legend of a man. In his book he addresses the question of being scared or frightened. Lord knows he had many scary encounters and situations throughout his career. He said he did not equate scary situations with fright. Basically he said being momentarily scared, startled, or whatever, was in his opinion, a natural reaction to danger. He stated apprehension, conquered and mastered, is quite different from fear that debilitates. One last comment on this - I have to say that LB II was indeed a defining moment in my life as it was a true test of my metal as to what I would do when presented with a very dangerous situation. I am very proud I acted appropriately and kept my cool under fire.

So our dinner that evening of the 21st was rather short and light on food. If I recall correctly, soup and saltine crackers was about it. I just could not stomach too much and as I looked around the dining room I saw I was not alone in my loss of appetite.

What I remember differently about the mission briefing was that SAC finally relented and the overall combat tactics were modified. Our target for the mission was the Bac Mai Storage Facilities located several nautical miles south of Hanoi. Everything was briefed in the usual order, ending with the Chaplain's blessing. The crews broke up for our specialized sessions and again nothing seemed out of the ordinary. That is not to infer these missions were anything but ordinary. It just means the tactics for us EW's pretty much remained unchanged. I considered my gunner and I were a really solid defensive team. We continually discussed several possible scenarios and what we would do if they occurred, especially if we were ever attacked by a MiG. I did take a few minutes to review the SAM evasive maneuver procedure with the pilots and grabbed a couple of bags of peanuts from the hooch outside mission operations before heading off to the Comm Shop and Life-Support to pick up survival vests and our pop-guns. Finally we headed to our aircraft.

As I said earlier, as a crew we were very senior and had often been the cell lead aircraft on missions. We thought we might be the lead that night, but another crew for the mission was very senior in rank to us so they were tagged for the role.

I usually say rank-had-it's-privilege, although on that particular night, being "out ranked" probably saved our lives as you will read shortly. A typical B-52 Arc Light mission consisted of a single three-ship cell, but LB II used a wave of many three-ship cells, each cell consisting of three B-52s in trail separated by ½ mile. So for this wave we were assigned to be the number two aircraft in Blue Cell - aka Blue 2. The Bach Mai storage warehouses were new targets located just to the south of Hanoi and in very close proximity to Bac Mai airfield and hospital. My aircraft preflight was normal; except for giving my ejection seat a very special, detailed and loving check to make sure it would not fail me if I needed it. Taxi and takeoff were normal, if anything could be called normal about a massive number of BUFFs moving around in the dark, and we departed U-Tapao around 2 am local. Our three-ship cell was part of the much larger wave of B-52s of maybe 60 or more aircraft. We were all heading to the Hanoi-Haiphong Route Pack 6 area to attack a wide variety of targets. There were a total of six B-52s heading to our specific target area. Our cell was 3-4 minutes behind the three-ship "Scarlet" cell. We proceeded north to the target area and as we flew deeper into North Vietnam, I started picking up the usual very heavy level of radar activity and commenced with some of my electronic countermeasures. We could hear all the SAM and MiG (bandit) warnings being called by the Navy ship, with the code name of Red Crown, sitting in the Gulf of Tonkin monitoring the air war. Since our B-52s and support aircraft of USAF and Navy F-4's, EA-6B's Navy jamming aircraft, and USAF EB-66s (also jamming aircraft) were so effective in distorting the SAM operators screens and suppressing SAM launches, the North Vietnamese would send up fighters - not to attack us, but to try to determine at what altitude we were flying. The fighters would pass that information back to their SAM operators to get them to set some of their missiles to detonate at our altitude when fired in a ballistic mode. We continued to press on to our target area and about five minutes prior to getting there my equipment started displaying SAM target trackers and uplink guidance radars. Much later our crew determined that somewhere between our Initial Point (IP) and our target, we had 8-10 plus SAMs shot at our cell, many of them in the air at the same time.

Subsequent to the mission we often discussed just how many SAMs we thought were fired at our cell, since in most cases it was pretty much impossible to tell if the SAM had your name on it or not. Sometimes you could tell, but not usually. That night we were pretty sure of our SAM count; at least during the short time between the IP and target. (This was later confirmed by the Blue 1 crew in another book). We began our evasive maneuvers and continued our track to the target. About then we heard a call over Guard from the Scarlet 3 co-pilot who was about four minutes ahead of us and headed to the same target. They had been hit by a SAM and were burning. We knew the crew well since they were from our home base. We continued pressing on to our target while continuing to evade what seemed like an endless number of SAMs. About two minutes from the target, we lost contact with our lead aircraft which was no longer responding to calls from our aircraft. Also about that time, I began to see their tail gunner's fire control radar on my warning receiver. The signal was moving down at my two o'clock position and eventually completely disappeared from my warning receiver. Shortly after that we started to pick up parachute beepers on guard frequency. Our parachutes were equipped with emergency beepers that activated when the parachutes opened. The strength of the beepers was a further indication to us that Blue 1 had likely been hit and the crew had bailed out. At one point the beepers got so loud (as we were close to Blue1) we had to turn our guard radio off since we could no longer hear the threat warning calls being made. Guard radio let you monitor all radio communications from all aircraft, ships, etc. which broadcasted on specific frequencies of 121.5 Mhz and 243.0 Mhz.

Bach Mai storage warehouses damage.

It was then our Pilot called Blue 3 behind us to tell them we suspected we lost Blue 1 and we would assume the role of cell lead, but keep the call sign of Blue 2. After what seemed a lifetime of flying straight-and-level, we finally made it to our target, dropped our bombs, made our post-target-turn (PTT), and once again began evasive maneuvers to avoid the constant barrage of SAMs. I do remember thinking, "Are we ever going to get out of range of those damn missiles?" I'd also like to mention I think having a gunner sitting in the tail of the aircraft was a lifesaver. Countless times he called out the track of SAMs coming at us and told the Pilot when to break and in which direction. He was something special; he told us when we got back to our trailer and after a few of post-mission drinks that a SAM had gone between the wing and fuselage but detonated above us. God was truly our 7th crewmember that night. As the cells approached the coast, Red Crown always polled the crews coming out to see how many made it. We heard the call by Scarlet cell that they had two aircraft but had lost Scarlet 3. Red Crown polled our cell and our Pilot told them that we also had two and had lost Blue 1.

So we made it back to U-Tapao from what turned out to be our worst mission. It wasn't until we landed that our pilot told the rest of the crew we had some damage to our left tip tank and wing, but again we lucked out and made it home safely. We had lost Scarlet 3 and Blue 1, both in front of us. We did hit the target and did grave damage to it. (See BDA picture.) Three of the crew on Scarlet 3 were killed and the other three were POWs. All six crew members from our lead aircraft, Blue 1 made it out of their aircraft and became POWs. All were repatriated in March of 1973.

It wasn't until years later I realized just how lucky our crew was to make it home. I did so while reflecting back on this mission while listening to the mission recording we made. Do I have any idea of the total number of SAMs fired at our cell? No, but it was many more than the original 10 I estimated. And I have on occasion wondered why two of our six aircraft were shot down. Luck? Maybe. But the North Vietnamese SAM crews were damn good after so many years of experience. I think having only six of us heading for that target may have reduced the level of ECM effectiveness and could have been a contributing factor in the shoot-downs. We will never know for sure. But, there is no doubt in my mind, that no matter how well all our crews did their job, Lady Luck was indeed a player in the outcome. It was a close call for sure.

I went on to fly my fifth and last LB II mission on 26 December when the night's target was SAM site VN-549 - the site suspected of shooting down a number of B-52s. We rotated to Guam late the next day and only sat as a spare on the last night of LB II.

And as Forest Gump said, "That is all I have to say about that!"

The Infamous Linebacker II Routes
Fred Miranda

I was never a Crewdog, however, as an Aircraft Maintenance Officer, I always considered my mission to be "Keeping Crewdogs Alive!"

In December 1972, I was on the SAC Headquarters DCS Logistics Staff, at Offutt AFB. I was a mid-grade Captain at the time. My office symbol was LGMMX (LGMM Being Aircraft Maintenance Management and the X denoting planning.) I was on my fifth PCS following tours at the 499th Air Refueling Wing at Westover, a year at Ton Son Nhut, Saigon, a 15-month tour at 2AF Headquarters, Barksdale, and two years at 3d Air Division/8th Air Force Headquarters on Guam. I guess you could say I was becoming a bit of a "Headquarters Weenie."

Because of the Guam experience I was selected to be the SAC LG Singlepoint for Arc Light and Young Tiger operations - later expanded to include, Bullet Shot, Constant Guard and eventually Linebacker I and II. I had help. Capt Jere Miller (later Colonel Miller) and CMsgt Ken Cousino were two individuals who come to mind as being almost completely involved in the support of SAC operations in SEASIA. We were busy. Not only did we have Arc Light and Young Tiger to plan for, but we also had the Single Integrated Operations Plan (SIOP) and SAC day-to-day operations. We worked more than an eight-hour day and Saturday mornings were routine until Gen Meyer showed up as CINCSAC. (He stopped coming to work on Saturday and soon the rest of the staff did too.)

There were other logistics sections involved as well: munitions kept close watch on the supply of Mark 82 and 117 bombs. Avionics monitored improvements to Bomb Nav and ECM systems. Airplane movements between WESTPAC and the CONUS were closely scheduled to distribute flying time as equally as possible among aircraft and maintain the Depot Maintenance Schedule.

When I arrived at SAC Headquarters, Maj Gen Pete Sianis was the DCS Logistics. General Sianis endorsed at least two of my OERs during that time and I always hesitated to tell what I knew about the Linebacker II planning because I thought I was knocking someone who was key to my promotions. Eventually I decided I had to share what I witnessed before all of those who were involved and who care were gone.

It was a Friday evening, 14 December 72. I had been home after dinner with the family and had finished a couple of drinks. The phone rang about 1930. It was Col Gorden Krentz, my boss's-boss's-boss! Col Krentz was the SAC LGM, Director of Aircraft Maintenance. He had also been my first DCM at Westover. He asked me to come to work. I asked him if I needed to come in uniform and he said, "Yes!"

I reported to Building 500 in about 30 minutes. The "Maximum Effort" message from Adm Morrer had been received and we had to get a briefing together for Gen Meyer ASAP. As I recall the briefing took place at 0200 local time on Saturday morning. I presented the DCS Logistics portion of the briefing which covered numbers of aircraft B-52D and B-52G and where they would come from, Anderson and U-Tapao. Routing was not covered during that briefing. The DCS Operations Planners were very busy coordinating with 8th Air Force and developing the required information for the actual missions themselves.

The rest of this story is adapted from a letter I wrote to the editor of *Air Force Magazine* which appeared in the December 2015 issue. The letter was in response to an earlier article entitled, *"The Nightmare Before Christmas"*.

Maj Gen Sianis had moved to the DCS Operations position. Maj Gen George McKee was the SAC DCS Logistics during Linebacker II.

It was late Saturday morning when the SAC Contingency Operations staff was about to brief Maj Gen Pete Sianis, SAC Deputy

Chief of Staff for Operations (DO), on the plan they had developed - I was there. We were in the DO's outer office on the second level of the SAC Underground Command Post waiting to go in. I was the SAC DCS/Logistics' representative. The route charts showed several different routes leading to Hanoi. The Andersen Air Force Base aircraft route led from Guam to an air refueling area north of Luzon, and then to Point Juliet in the South China Sea, and then northwest to Hanoi. I do not recall the U-Tapao routes, but there was more than one.

Maj Gen Sianis walked out of his inner office while we were waiting, took a look at the map, and said, "That's not the way we do it!" Then he removed the colored tape showing the Andersen B-52 routing from the map and rerouted that bomber stream to a route over South Vietnam into Laos and forming up with the U-Tapao bomber stream. He also changed the post-target exit routing to one requiring all aircraft to make a right turn after dropping bombs and stated, "One way in and one way out!" He then instructed his staff to go make those changes and come back with the briefing. I will never forget how the map looked after Maj Gen Sianis made changes. The colored tape was hanging loosely and the general made a comment, "You guys probably have a lot of this tape, don't you?"

I recall at the time being astounded when Maj Gen Sianis said, "That's not the way we do it!" I thought, "Who's this 'we'?" SAC had never done this before.

This was a significant last-minute change resulting in re-planning additional post-strike refueling, and the now infamous "post-target turn." He essentially took the planning function away from the majors and lieutenant colonels and straitjacketed them with the "One way in - one way out" directive. No one questioned the SAC DCS/ Operations. The CINCSAC, Gen John C. Meyer, was a TAC guy. It took three days and some real heroics by people like Brig Gen Glenn R. Sullivan at U-Tapao to affect change to this faulty planning.

Maj Gen Sianis was a WWII B-24/B17 pilot and a Squadron Commander. He had flown 20 combat missions from 8th Air Force bases in England and accumulated 150 combat flying hours. I'm thinking that was the "we" he was talking about.

I was in the SAC Underground Command Post "Battle Cab" with Maj Gen McKee and the rest of the Senior Staff for all of the Linebacker II Time Over Target (TOT) times and had a clear view of Maj Gen Sianis during days 1, 2 and 3. You could almost see him sweat and get smaller as B-52s were shot out of the sky during the post target turn. It's something I will never forget.

The Following Obituary was published in *The New York Times* on February 10, 1988.

Pete C. Sianis, Hanoi Raids' Planner, 68

BELLEVUE, Neb., Feb. 9— Maj. Gen. Pete C. Sianis, retired, an Air Force officer who planned devastating B-52 bombing raids on North Vietnam, died Saturday after a long illness. He was 68 years old and lived in Bellevue.

As the Strategic Air Command's deputy chief of staff for operations, General Sianis was the architect of an operation in which B-52's bombed North Vietnam's capital, Hanoi, and its principal port, Haiphong, from Dec. 18 to Dec. 29, 1972, after negotiations to end the Vietnam War had broken off.

56-0589 at Sheppard Air Park

It Was Broken Before We Took Off

Randy Wooten

Most war stories begin where others end. It makes sense if you accept the concept that wars are a continuous series of events, written on the world stage, where we all serve only as "bit actors" in an ongoing saga.

On April 24, 1972, Captain John Alward and his crew recovered a battle damaged B-52D into Da Nang, South Vietnam. His aircraft had been hit on the target run into Vinh, North Vietnam, and had two engines out and only partial power on the other two on the right wing. The situation set the crew up to have to make a "three or four engines out on one side" landing.

The weather was IFR at Da Nang and the crew made an ILS approach. Because of poor instrumentation and gross weight info for the landing data, they made a landing attempt, and go-around - which in itself is a great war story. They finally got it on the ground at Da Nang, were picked-up, and returned to U-Tapao to tell their harrowing adventure. This closed their chapter, and began mine.

I was the co-pilot on Dyess crew S-19, under the command of Lt Col Adam Mizinski, flying out of U-Tapao. We had flown on 23 April, and were DNIF cover that next day. Early in the morning, Adam came into my crew trailer quarters and woke me up. He told me to get my flying gear, bring my performance manual and Dash-1; we were about to catch a KC-135 to Da Nang to bring back a wounded bird. We would be back in U-Tapao by the afternoon.

We were a "minimum crew" for the B-52—two pilots, the nav, and a gunner. The gunner for our recovery mission was a spare, since our gunner had been called to cover another DNIF gunner earlier in the morning. At this point in my story, I'll offer an apology to Col. Mizinski.

A few years after this event, he wrote a story of our adventure, and I remember reading it and thinking, "That's not exactly the way I remember it." Nevertheless, my story will differ slightly from his version because I will tell it as I viewed it through my eyes, which has become "my" memory of those same events. All the big pieces are the same, just some of the details are a little different - so, sorry boss - please forgive the co-pilot for the error of his ways.

We arrived at Da Nang and went immediately to see the airplane, since we were going to have to "fly it out" that day. I don't think so! It was a sad sight - sitting there with an army of technicians covering it like flies, with fuel dripping from every place you'd expect, and several that you knew were "just wrong." We finally found the guy in charge, and started asking him questions. The plane (B-52D, 56-0589) had over 400 new openings in it, which were not approved by Boeing in its original design. The number I was given later was 427 holes.

We also quickly found there was no way we were getting the plane out that day, and were told, "Come back tomorrow and we'll see." We did.

We did a preflight, after the boss checked the forms, which was as you'd expect - a sea of red X's. It took a while to hear all the things that didn't work, like pressurization, multiple warning lights, and several fuel transfer valves. Half of the firewall T-handles were safety-wired open, as well. The cockpit looked like a simulator at the end of a three-hour lesson - a sea of red and yellow warning and caution lights.

What did work were the engines, some hydraulics, and radios. We were going to be led back to U-Tapao by flying in formation, gear down, with our KC-135 on a day-only mission. At that time, there were reports of SAMs being moved into Laos and northern Cambodia, so we were instructed to fly down the Vietnamese coast and then turn west over southern Cambodia and on to U-Tapao.

We mounted up into our crew positions and began engine start procedures. The poor nav had a slow but steady drip of JP-4 falling onto his crew position, which could be smelled throughout the aircraft cabin. We finally got six, and then seven engines started, but no matter what we did, we couldn't get the last one started. Later we found it was because of an inoperative starter valve. So, since we had broken every other peacetime safety procedure, why not add a seven-engine takeoff to our list of offenses?

Adam told me to get out the performance manual and start computing seven-engine data. I'd never visited those particular pages before. I worked up the data, and the boss talked me through the techniques we would use during take-off. The one I remember vividly was his instructions to put in three degrees of rudder trim at a time if he nodded his head and said "Rudder." I don't know if that is ever discussed or even captured in any B-52 official or "hot-tips" pamphlet. I've never seen or heard any reference to it since that day. But, I did know that Adam Mizinski was the "King" as far as I was concerned, and if that's what he wanted, that's what I'd give him.

Anyway, it didn't happen. We taxied with seven engines to the hold line, and then got a call from the command post to return to parking. The pucker-factor was just too great for the staff guys at HQ, so we turned the bird back over to maintenance, with plans to try again on day three of the adventure.

Day three was much like day two. We got seven engines started and the eighth still wouldn't rotate. They decided to take a starter control valve off a running engine and put it on the inoperable engine to start it so we'd have eight engines. However, the engine with the missing starter control valve would leave a huge "bleed air leak" in the nacelle that the Dash-1referred to as "an impending disaster." But we did anyway and had eight engines running and we left.

We followed modified checklist protocols, and came to where we were directed to close the cross feed fuel control valves. I did, and sure enough all the engines on the right side started to unwind. The procedures called for the co-pilot to keep his fingers on the valves for such eventualities. We caught the engines before they reached 60%, but the airplane made a hell-of-a-yaw.

We're following our KC-135 at about 23,000 feet, and came to the time we needed to open a fuel tank to transfer fuel. It wouldn't work. That meant we had trapped fuel we could not use, which was going to impact our range. After some quick computations, we notified the KC-135 we needed to go direct to U-Tapao. We felt that was a better choice of routes regardless of the potential SAM threat. Besides, with no EW, we wouldn't know about the SAMs anyway.

We made the course change, and as we were approaching the field we found there were thunderstorms in the area. We delayed our descent, and then tried to go to airbrakes six to get down to pattern altitude. There were no airbrakes. So we ended up doing 360s to enter the pattern. The boss put it down ever so gracefully, and we already knew the drag chute was inoperative. We rolled out with fire equipment all over the place and stopped on the end of the runway and dismounted. We proceeded to maintenance debriefing, for what it was worth. When every system is broken it doesn't take long.

I did a quick fuel reading, and subtracted the trapped fuel. We had enough fuel for another five minutes before flame out. My flight records show I logged 1.8 hours of co-pilot VFR time in B-52D, 589, on April 27, 1972.

A B-52 lands at U-T.

They towed 589 to the hangar, and it took until Jan 9, 1973, before it was back in commission. It was eventually sent to Sheppard AFB, Texas, as a maintenance trainer, and resided in their Air Park until approximately 2010. I say "approximately" because that's as close as the folks I talked to at Sheppard AFB could come to the actual date. Even though it had survived the battle damage and the war, it was eventually scraped due to severe corrosion and a G-model took its place on display.

As a side note, I was assigned to SAC HQ in the plans directorate, and in 1985-88 wrote and executed the program plan to retire all the B-52Ds, as part of the B-1B bed-down. We had 25 requests for static display aircraft from around the world, to include Guam, England, and South Korea. Since I wrote the Project Plan, I selected tail numbers that matched my flight records. So whenever I see a “D” on display, I can proudly say, “I flew that airplane.”

Arc Light A to Z

Ted Lesher

Many different accounts and even entire books have been written about the B-52 operations during the Vietnam War. I don't want to rehash this well-covered topic, but I think I can add a few stories not told elsewhere. Arc Light was the primary code name assigned to most of this activity, but other offshoots such as Bullet Shot, Hot Tip, Linebacker, and probably a few others also apply. I would like to stake a claim as perhaps the only crewdog there at the very beginning and the very end of Arc Light, as well as a good many points in between.

Arc Light Before It Even Had a Name

I entered the Air Force in late 1961 and for more than two years went through the high-pressure training it took to get from being a civilian off the streets to occupying a crew position in the B-52. By early 1964 I had finished the Electronic Warfare School at Mather AFB, near Sacramento, California, and departed for my first operational assignment in the 320th Bomb Wing on the other side of the base. Life in the real Air Force proved to be much different than the training for it, and by mid-1964 I was pulling a lot of unbelievably boring nuclear alert sessions. Fortunately, about that time some things were kicking up promising to make the job a whole lot more interesting.

Trouble in Vietnam had been brewing for a long time and by mid-1964 it looked like it was going to turn into a real war. The Strategic Air Command (SAC) resisted diversion from its nuclear mission, but higher powers prevailed and our F-models at Mather became a test bed for B-52 conventional bombing equipment and tactics. I don't know if conventional high-explosive bombs had ever been dropped from B-52s before, but our unit became deeply engaged in the process. At the time, the B-52's bomb bay would only accommodate 27 conventional bombs and Boeing was contracted to find a way to mount more on the wings. We successfully tested the resulting external racks, which practically doubled the available bomb load. We made live drops of BLU-3 cluster bombs, a diabolical weapon designed to detonate thousands of bomblets on impact and shoot millions of little ball bearings in all directions over a wide area. We practiced three-ship close-formation flying, which may or may not have ever been done in the B-52s before, as well as looser cell formation which later became the mainstay of conventional operations.

I was living in the grossly misnamed "Sin City," an on-base enclave of tract homes which served as Mather's bachelor officer quarters. One house was occupied by three female officers, and all the rest by young guys like me - mostly students from the Air Training Command side of the base. A certain amount of partying went on there but it was pretty tame by present-day standards. Off-base living was about the same. In Sacramento the hottest place to go on Friday night was a restaurant called Schiedel's Bavaria, which featured a German accordion player named Duane Boderman who sang songs like *Goodnight Irene* at the bar. That's where the guys went to meet the girls. As I look back, that seems like it was not only a different time but a different universe.

Toward the end of 1964 it started looking like our conventional bombing activity was going to be more than a test and we might actually have to use it. In December the wing stood down from our nuclear alert and the crews were sent home with instructions to pack our bags for 30 days, stay within three rings of the telephone, and be ready to deploy to the island of Guam. This standby order blew over after a few days, but happened again about a month later, when it blew over again.

Deployment Number One

February - June 1965

As the saying goes, the third time's a charm. On February 7, 1965, the crews were called in and briefed for a massive mission to be flown against Phuc Yen airfield near Hanoi. Once again they were sent home on and placed on telephone alert and a few days later the call actually came. About 0600 on February 11th I reported to the alert facility with the gear I'd need to be gone for 30 days, which I guess is how long SAC thought it would take to end the war. We gathered in the normal alert briefing room, picked up our mission materials, and were briefed. All 15 Mather AFB bombers were bound for Guam, along with a like number of tankers that would stop in Hawaii en-route. A similar formation from Barksdale AFB, Louisiana, would follow. My aircraft was parked in the alert Christmas tree just outside the briefing room. We went outside and did a preflight much like we did for any other flight, and I think this operation must have been pretty well planned. I recall there was very little confusion, particularly considering how many aircraft were involved, and how fast things were happening.

We started engines around 0800 and got lined up for takeoff, with some aircraft coming from the alert Christmas tree and others from the normal parking ramp. As we taxied out, some famous words were broadcast from the command post: "The nickname for this operation is Arc Light." That was the first time any of us had ever heard the term. Little did we know eight years later it would still be going.

We took off, headed west, and had a top-off aerial refueling shortly after reaching cruising altitude. It was called a "Mass Gas" formation, with multiple bombers in trail on multiple parallel tracks, and 30 sets of contrails heading west above the Golden Gate. I later heard it scared the pants off of a good many citizens of San Francisco who thought the end of the world had come. For us the trip was quite uneventful, and before sundown we were checking into the BOQ at Andersen AFB, Guam. We were billeted in a large multi-story concrete building - something like a college dorm with each six-man crew occupying two rooms with a shared bath.

Our original plan was to spend a couple of days on Guam and then fly a mass attack against Phuc Yen Airfield, but for some unknown reason it never happened. Instead, what followed was the most bizarre period in my life - a period I have come to think of as "The Phony War." For the first few days we were basically on full-time alert, which was considerably reduced when it became apparent nothing was going to happen anytime soon. The powers that be figured if we ever attacked it would be at night, so they kept us on alert during the days but released us when it got too late to hit Vietnam in the dark. The result was a mass exodus to the Officers' Club by the crews around sundown every evening. We had some very memorable games of liar's dice, bar talk of all sorts, and the spinning of some great war stories. I had always been uneasy around the crew force of the time, which consisted of many senior officers dating back to World War II - mixed in with a few new arrivals like myself and some others in between. On Guam we simply bonded into a giant group of aviators, resulting in me feeling very comfortable around crewdogs of any vintage ever since.

After a few weeks the staff recognized that we still needed to maintain flying proficiency, so we started flying training missions. By stateside standards these missions were pretty short: consisting of air refueling, a little simulated bombing, and some takeoff and landing pattern work. We even did some low-level terrain avoidance coming down the Marianas chain, in which we found the only terrain for hundreds of miles around and then avoided it.

We also started getting entire days off. When those occurred, many of us felt a need to venture off base, but transportation was a problem. Like several others I bought a motorcycle and rode it all around the island. There was an abandoned WWII airfield nearby with a couple of 8,000-foot runways and a friend of mine with a similar machine and I would go out there and race. We had fairly big bikes that could get close to 100 MPH. Forty years later I encountered the same guy at a reunion, where we were both surprised to remarkably still be in one piece.

Eventually, like all good things, the party on Guam had to end. After being there for four months, the first mission was originally scheduled to launch on June 15, but was delayed when Super Typhoon Dinah popped up on the scheduled route. On the 17th the typhoon hit 160 knots and was centered right in the refueling area north of the Philippines, but it was predicted to rapidly move off. The next day, June 18, 1965, we finally flew Arc Light 1. That particular mission has been endlessly documented and I'll avoid going over all the things that have been covered so well before. I will note the mission was nothing like the original Phuc Yen plan, but instead targeted a patch of jungle in South Vietnam not far from Saigon.

We had ten cells of three bombers each, with each cell doing its own independent navigation, everything done under radio silence, with the meet-up with the tankers based solely on timing. These were the days long before GPS and weather satellites, and inflight we encountered 200-knot tailwinds instead of the forecast 100-knot headwinds. This resulted in timing being all messed up and nobody knowing where anyone else was.

I was flying in Blue 2, the second aircraft in the second cell, making us the fifth aircraft inbound on Arc Light 1. As we were approaching the refueling area, our gunner called from his crew station in the tail of the aircraft saying "There's a bright light back there." A few seconds later he called out "That's really a bright light!" About then we started hearing emergency locater beacons going off on guard frequency. It wasn't hard to figure out an aircraft had blown up and the crew was bailing out. A while later, silence was broken on the long-distance High Frequency short-wave radio I monitored, with a Zippo Lost Aircraft report. I told the crew to listen in.

Another lost aircraft report followed, and it was evident two bombers must have collided. Our squadron ops officer, who was on board our aircraft as an observer, identified the crews. It was a mystery to us what had happened since the two B-52s involved were flying in separate cells that should have been many miles apart.

The remainder of the mission was uneventful. About ten minutes before reaching the target a voice came booming over the emergency frequency announcing "Heavy artillery from ground to 30,000 feet" at such-and-such location. Our navigator exclaimed "That's where we're going!" We had a discussion about it, and finally concluded the heavy artillery was probably us. It was the first time the B-52 crews had heard an artillery warning or even knew such a thing existed. You'd think they could have told us about it first. After we left the target area, another unidentified voice, presumably someone doing recon down low commented "That's good bombing." I thought to myself "I should hope so – after all, we were SAC, these were B-52s, and bombing is what we did, even though this was the first time ever for the real thing." But as it turned out, only about half the bombs were on target and there was negligible effect on the target.

Six hours later we were approaching Guam with 26 bombers to land. We started out with 30, two collided and two diverted elsewhere to land. The recovery plan, like many aspects of the mission, was unnecessarily complicated and could easily have led to more trouble. It appeared to be based on the way civilian air traffic of the time was handled, with stacks of aircraft circling in holding patterns until they could peel off one by one for landing.

There were multiple stacks of aircraft - I don't remember the exact configuration but there were something like ten of them. There was a stack for each cell, with the three aircraft in each cell circling at different altitudes within their stack. There were airplanes going different directions everywhere and somehow all of them got down without colliding or running out of gas.

At debriefing we found out what had happened to the two lost aircraft: the unexpected winds had seriously thrown off the planned timing for the air refueling rendezvous and radio silence made it impossible to coordinate a response. The leading cell elected to do a 360 degree turn to get back on time, and ended up flying head-on through the following cell. The mission had been intensely monitored all the way up to presidential level and we had a disaster before we even got halfway to the target. At debriefing the crews were met by a very glum Brig Gen Harold Ohlke, 3rd Air Division Commander and the senior officer at Andersen. I suspect he had little, if anything, to do with the planning of Arc Light 1, since such a high priority operation typically would have come directly from SAC headquarters. Practically all of our problems originated at that level, but Gen Ohlke was nonetheless relieved of his command within a few days.

It was nearly a month before any new Arc Light missions were flown. In the interim, temporary duty (TDY) time was running out for this first batch of Arc Light deployers and we started returning to our home stations. My crew was among the first to go, about 10 days after Arc Light 1. After several delays we finally got off the ground about midnight - a single-ship with no air refueling with a direct shot to Sacramento passing near Wake and Midway islands on the way. We ferried our B-52 home loaded with cargo, with plywood platforms installed in the bomb bay and the 47 section aft of the rear wheel well stuffed with cargo nets. We carried the belongings of the crew guys that hadn't survived the collision, seven motorcycles (including my own), and a motley assortment of all kinds of things that somehow made it on board.

We had trouble right off the bat: a tip gear wouldn't retract, and we went into a brief holding pattern while consulting with the command post to figure out what to do about it. The consensus was that even with the increased drag we still had enough gas to make it, and the last words from Guam were for us to call for a strip alert tanker at the other end if it was needed. "Over and out."

It would be a 12-hour flight with a decision point about nine hours after takeoff. Before that point we could divert into Hawaii if needed and afterwards the mainland was closer. At the decision point we would be three hours from anywhere. The whole crew was engaged in handling the fuel situation. In those days there wasn't much equipment to navigate over open water, so I shot celestial fixes all the way across. Celestial is pretty accurate when the stars are out but becomes much less so when the sun comes up, which happened after about four hours into our flight. The moon came up later and gave us something else to help us navigate, but our position was still approximate. The nav team kept close watch on ETA and the pilots on our fuel, and when the decision time came we pressed on towards home.

Gradually things started looking worse and about an hour and a half off the coast our pilot decided it would be prudent to follow the advice from Guam and call for a strip alert tanker. We were well out of UHF radio range of any base, and as custodian of the long-range HF radio it was my job to make the radio call. SAC maintained its own HF network, called Giant Talk, which had a number of ground stations operating on a handful of fixed frequencies. Crew members of the day were aware of terms such as Democrat, Sky King, and 4725 Upper.

I attempted to make contact, but the HF radio was known for being temperamental and it picked that moment to act up. I tried every Giant Talk frequency and every time I keyed the mike it would tune continuously. By then the crew was getting nervous.

There was a second HF network used by the rest of the Air Force and, somewhat in desperation, I gave it a try. Much to my surprise, I tuned in McClellan AFB on 11176 and it stayed there. I asked for a phone patch to the SAC Command Post at Mather. I don't think they handled that kind of thing very often because it took them a while to get us hooked up. When we got through, I told the command post what was going on and could hear the alarm in the controller's response. About 30 seconds later a voice like Jehovah came booming in over the radio. This turned out to be the 15th Air Force Command Post, and they got a KC-135 launched out of Travis for us in very short order.

We were still out of UHF range but could hear the tanker check in on HF with McClellan. I didn't dare change frequency and we ended up talking directly to the tanker on HF while McClellan cleared everything else off the channel. Our nav team set up a point-parallel rendezvous, in which the tanker approaches head-on but offset to one side. Normally our radar nav would see the tanker coming and at the appropriate range direct a turn; if all went well, the tanker would do a 180 and roll out a couple of miles in front of us, headed in the same direction and ready to refuel. That's exactly what happened, except after the turn there wasn't any tanker.

It took a while to sort out what was going on. Without getting too technical, the B-52's radar has an inherent range ambiguity so that a sufficiently bright target at maybe 120 miles could show up on the screen at 20 miles. That's where the tanker was - still 100 miles ahead. It took another 15 minutes or so to get turned around to make up the distance and do another rendezvous. In the meanwhile our fuel was getting very low and our pilot was getting very nervous. While we were closing in on the tanker he muttered on interphone, "This is going to be just like Yuba City." He was referring to an incident where Mather had lost a B-52 under similar circumstances just four years before. With the pressure on, he was initially a little jumpy and had trouble making contact with the boom of the tanker, but then he settled down and between him and the boom operator they successfully transferred the fuel and saved the ship.

An hour later, just about dusk, after a very long haul from Guam, we were home and on the ground. We may have been the first returning crew from the first B-52 combat mission in history. We had been through something of an ordeal on the way back, and thought there might be a pat on the back for us there somewhere. It turned out customs was the big deal and they wouldn't let us go anywhere until they had unloaded and checked all our cargo, which took hours. So much for the pat on the back! That's why they call us crewdogs.

Deployment Number Two

November 1965 – March 1966

Five months later, around the end of November of 1965, the 320th Bomb Wing at Mather redeployed to Guam. There was a huge difference between the last deployment and that one. SAC had got its act together and worked up a practical operations plan and Arc Light missions were routine by then. There was a red-jacketed classified manual that was essentially a catalog of Arc Light errors and how to avoid them. (It got thicker and thicker through the years.) There was even a school for aircrews – SAC Contingency Aircrew Training (SCAT) – and a mini-check ride on a crew's first flight. We started school the day after we arrived. The first lesson was "Follow the leader." This was so different from our previous Arc Light experience we couldn't quite believe it. Our instructor emphasized the point by putting his hands out like little airplanes in classic aviator fashion, and marching around the room with one little airplane following the other wherever it might go.

I flew 28 combat missions on that tour, all of them out of Guam and all of them unremarkable. It was the ancillary activities which were interesting.

Instead of a motorcycle, I got a "Guam Bomb," a junk car I acquired from another crewdog who was going home. It was a four-door Pontiac that was thoroughly rusted out like every other car more than a few years old on the island. The left rear door wouldn't open, the right rear window wouldn't close, the shotgun seat had settled through the floorboards and was propped up with a wooden slat, and I carried several cans of oil to replace what was going out the tailpipe. But, as they said, any car on Guam that would run was worth at least $20, and that's what I paid for it. Remarkably the powers on the base didn't have any problem with those things.

There was an existing rest-and-recuperation (R&R) program at Andersen, mostly geared toward permanent-party families who understandably got island fever being stuck there for an entire PCS tour. It operated by means of the so-called "Hong Kong Flyer," a C-97 making utility circuits to Yokota, Taipei, and Hong Kong. Arc Light crews found out about the flights and about two weeks after we arrived on Guam we went to Japan to recuperate. I took off from Yokota and went to Tokyo on my own, taking the train into town and ended up at the New Otani Hotel. It was an upscale establishment and, like the rest of Japan, was dirt cheap at the time. That was the first time I had set foot in a foreign country and one of many times in my military career when I found myself a bit amazed at being there at all.

About a month later there was a mass gathering of news reporters on Guam trying to fly a B-52 combat mission. We drew R.W. Apple of *The New York Times,* who had previously been embedded with the Army in Vietnam and now became embedded with our crew for a few days. He had some war stories about getting shot at and diving into rice paddies for cover, but after the mission he commented that air refueling was the most frightening thing he had ever had experienced.

He occupied the instructor pilot's seat which was directly under the air refueling receptacle. During the initial contact with the tanker there's a big "clunk" when the boom plugs in and the toggles engage. I guess we should have told him about that. Afterwards we had some beers at Gilligan's Island and piled into my Guam Bomb for dinner at the Top o' the Mar, the Navy Officers' Club. The resulting article can be found in *The New York Times* archive for January 13, 1966.

Incidentally, I was present when Gilligan's Island was named. A roach coach and beer keg were stationed near the place where crews exited debriefing, providing very welcome refreshment after a 16-hour day. The commander of the provisional bomb wing, Col. William Cumiskey, was there one afternoon with a crowd of crewdogs and offhandedly commented "We ought to call it Gilligan's Island." Everybody laughed – one always laughs at the colonel's jokes. He was out there again a couple of days later after another mission, and in some irritation wanted to know where the sign was. What sign? Well, the Gilligan's Island sign, of course. And so it was called, for evermore.

On 19 February 1966 the crews were standing at attention in the Arc Light briefing facility as they escorted in a one-star general with a gray, somewhat non-regulation haircut. The briefing started "General Stewart, gentlemen ...." Until then we didn't know he was even on the island, but it was Jimmy Stewart, the actor and aviator. He flew the mission (with another crew, not mine) and afterward had a beer with us at Gilligan's Island. He evidently had been in Vietnam and I took the opportunity to ask him where he was headed next. His response was "How's that?" After all those years around airplanes it seemed that he was somewhat hard of hearing, something I can relate to myself as the years go by.

There were several experiments in defoliating Vietnam, including setting the jungle on fire. On 11 March 1966, I flew Hot Tip 1, consisting of six three-ship cells loaded with incendiary bombs. The story at the time was we started a huge fire which generated so much convection it created its own thunderstorm and put itself out. I don't know if that's true or not, but the whole idea was eventually abandoned.

Shortly before we were due to redeploy, I took my Guam Bomb down to Tarague Beach, an absolutely gorgeous facility on base property. There is a very steep section on the road which turned out to be too much for the old Pontiac and made the engine melt down and seize. We were able to scrounge an alert truck from somewhere and push the car up the hill and out the main gate and got someone to haul it away. That was the last interesting thing that happened on that tour.

Deployment Number Three

September 1967 – April 1968

In June of 1967 I departed Mather for good and arrived at Westover AFB, Massachusetts, which was just getting into the Arc Light business. I was a bachelor but Arc Light TDY was not only hard on many marriages, it was hard on my marriage prospects as well. I had a serious girlfriend in Sacramento, but our relationship had been interrupted by my second deployment and even when I was home it was hard to keep a date. She visited me in Massachusetts and we had a wonderful time in New York City and at Expo '67 in Montreal. When my third deployment came up she decided that was just the way it was going to be and she didn't want any part of it. We said goodbye for good, by phone, on the eve of my third departure to Guam.

That Arc Light tour was radically different. The Air Force had opened U-Tapao Royal Thai Navy Airfield (U-T) in Thailand and we were mostly operating out of there. It had numerous advantages and the missions were only about three hours long. The base was the "Wild West" compared to Guam. It wasn't swarming with O-6s and civilian dignitaries getting tickets punched, the Officers' Club wasn't dominated by colonels' wives, and there were girls - lots and lots of girls. It was "Bachelor Heaven!"

It also had a great commander, Col Alex Talmant, who was all business and the best friend the crewdogs ever had. His philosophy was well-trained junior officers and enlisted people, who did the work day after day, knew their jobs better than anyone else, and he would look out for the little guy and keep the higher-ups at bay. One day at mission briefing he asked if anyone had any questions, and one crewdog observed the crews hadn't received any mail for a while. The crew control officer, a lieutenant colonel, was in the back of the room, and Alex shouted, "Hey, Fat Boy, where's the mail?" Fat Boy said, "It didn't come in." Alex said, "We know that – where's it at?" Soon we started getting mail. He made Brig Gen, lived to be nearly 90, and had a successful retirement career.

I flew 68 combat missions on that tour, 44 from U-T and the rest out of Guam. Some of them were noteworthy. On 20 December 1967 we were on a bomb run just south of the DMZ when up popped a Fansong radar - the guidance system for the Soviet SA-2 surface-to-air missile. I announced to the crew "Hey, guys, we got one," put the divert word out on the radio and started jamming the signal. Our prime tactic on a routine mission such as that was to abandon the bomb run and get out of there. They launched some missiles at us, which went sailing by. In my entire career as an EW, ending with 7,500 hours in the B-52 and 222 combat missions, that was the only time I was shot at and one of the few times I operated the ECM equipment when it mattered. Everyone on the crew was awarded the Distinguished Flying Cross for the episode.

In March of 1968 we were flying out of U-T making daily runs to Khe Sanh. I know the situation was radically different on the ground, but from the crewdog perspective it was very impersonal and routine. Bombing was not very precise in the early days of Arc Light. The B-52 was a nuclear bomber and the bombing equipment was designed only to be close enough for that purpose. But, by 1967 the Air Force had come up with a better way by using ground-directed radar. For many years crews had trained and been evaluated through a ground precision radar scoring system which tracked the aircraft and predicted where a simulated bomb would hit. Combat Skyspot turned this procedure around, using the same precision radar to direct the aircraft to where to release the bombs. At first it didn't work too well – Col Talmant fulminated about all the wing commanders who had been fired because of bad bombs, and now the guys doing the scoring couldn't hit a barn door themselves – but the problems got worked out and became the standard procedure for the rest of the war. At Khe Sanh we were hitting very close to our own people, but aside from a heightened awareness it was pretty much business as usual for the crews.

With all the girls I, ahem, knew around U-Tapao, there was one young lady who seemed different from the others. Her name was Dang and she was a cashier at the Officers' Club. In the days before ATMs, the club was the place where every officer on base went to cash checks and she was consequently acquainted with everybody. There was a recreational facility on the beach near the end of the runway and one day I asked Dang if she would like to meet me there for dinner. She rather shyly agreed. That evening I waited and waited but she never showed up. When I confronted her later, she covered her face and said proper Thai girls couldn't be seen with guys like me. We remained friends but never met anywhere except at the cashier's cage. There's more to tell about Dang later.

I had a chance to do something few, if any, officer crewdogs have ever done: I flew two flights in the tail of a B-52. Our crew was assigned to fly a test mission out of Guam involving dropping a single bomb on a rock in the northern Marianas. There was absolutely nothing for the tail gunner or myself as the EW to do, so I asked our AC if we could switch places. We got the approval of our squadron ops officer and off we went. The lift-off over the cliff at Andersen is truly spectacular from the B-52 tail gunner's vantage point, and I really envy all the guys who had the best seat in the house in that incredible flying machine. This was a little surreptitious and the ground crew guy who opened the hatch after we landed was pretty surprised when he saw someone wearing captain's bars come climbing out. I was able to wrangle a second flight under similar circumstances, but that time when we taxied in there was a white-topped staff car with an O-6 inside looking us over. I had to dawdle in the tail until he went away, and that was the end of my gunnery career.

Our crew got another R&R to Japan, and so I headed solo into Tokyo and took the world's first bullet train to the historic city of Kyoto. That was 50 years ago and the U.S. still doesn't have a comparable train. I came back a couple of days later and spent my last night in the BOQ at Yokota. We were taking the C-97 flyer to Guam and were required to travel in uniform. I got up with plenty of time to spare and was getting dressed when I realized my uniform shirt was still hanging in a closet in Kyoto - 200 miles away. I dashed out to the Yokota BX and found it was closed for some reason that day, so I grabbed a taxi to Tachikawa AB, about five miles away, got a shirt there, and rushed back to the Yokota terminal. As I came in they were paging Capt Lesher. I ducked into the men's room and set up the uniform, hearing a few more increasingly urgent pages as I dressed. I ran out to the gate, which was just closing, to be met by my Aircraft Commander: "Where the hell have you been?"

That was the most exciting thing that happened on the tour.

Deployment Number 4

September 1968 – March 1969

We were back with the same crew from Westover and we were getting pretty good at those deployments. This time we traveled over in a KC-135 loaded with troops and cargo and had learned some tricks in making the normally uncomfortable journey. Troops rode sideways in long canvas benches attached to the fuselage on either side, along with cargo stacked up nearly to the ceiling in the middle of the aircraft. Aside from the total lack of wiggle room, the air would stratify a few hours out and it would be freezing cold at foot level. Those of us who knew what was coming would clamber out of our seats at the first opportunity and stake out some territory on top of the cargo, where we could stretch out and at least stay warm.

We had no R&R as such on the deployment, but an exchange program had been set up and we got to spend a week on the ground with the Army in Vietnam. We got qualified on the M-16 rifle, outfitted with fatigues in place of our usual flight suits, and with three bomber crews and three tanker crews took a C-97 out of Guam. We made the infamous Hong Kong approach into the old Kai Tak airport in Hong Kong harbor and stayed the night. That was my first trip to the amazing knockout of a place.

The next day it was on to Tan Son Nhut, where we were loaded onto a bus and taken to Saigon. I don't think we knew quite where we were going, but we unloaded by a high hedge with a passageway manned by armed guards where we were waved through. Somehow I was first in line, and when I turned the corner I was face to face with all four stars of Creighton Abrams. We had dinner there, outdoors on long picnic tables, just the general and the crewdogs. He was very affable, reminded me of Spencer Tracy, and got around to talk informally with everybody. I remember one thing he said, with some vehemence: "The higher you get the more they can chew on your fanny."

On the remainder of the trip we got an exciting low-level helicopter ride through a channel cut through the jungle for that purpose. We spent two nights in tents at Duc Co Special Forces Camp 30 miles west of Pleiku and visited a firebase where they were shelling a ridge in the distance. We also got escorted to a Montagnard village by Green Berets, and witnessed a B-52 strike a couple of miles away from a helicopter hovering at 2,000 feet. Even under those conditions the concussions were visceral.

I flew 74 combat missions on that tour, only one of which was memorable. There was a large-scale mission scheduled out of Guam, with a pretty full briefing room and the usual presentation by a multitude of staff officers. When the briefings concluded the division commander, Lt Gen Alvin Gillem, chased everyone out except the crews and himself. He announced "Here's where you are really going," and indicated a general area in Cambodia, which had been thoroughly off limits and actually still was. There were no charts to indicate the actual target and mission records were falsified. We were to be guided to the target from the ground by Combat Skyspot, go wherever they told us to go, and ignore any communications to the contrary. The chain of command for this operation evidently included a very few people between President Nixon and General Gillem. His parting words were "Don't f*** it up." We didn't, but maybe they did. One can research Operation Menu for details.

We flew a bomber home on a 17-hour nonstop flight from Guam to Westover - about 8,000 mile. We got some comments from air traffic controllers who in those days rarely saw flights that long. I accumulated 1,500 flying hours and 142 combat missions with that crew and would be remiss if I didn't mention them by name: John Dalton (P), Paul Daquisto (CP), Jack Simonfy (RN), Tom Radziewicz (N), and Charlie Marlow (G). They were the best bunch of crewdogs anyone could ever hope to be associated with.

Deployment Number 5

March - July 1970

I flew 51 more missions - 36 out of U-Tapao, 13 from Kadena AFB, Okinawa, and two from Guam. None of them were memorable in any way. I exchanged addresses with my friend Dang, the girl in the Officers' Club cashier's cage, and went home to a permanent change of station from Westover AFB, Massachusetts, to Beale AFB, California, just north of Sacramento.

Interlude

August 1970 – October 1973

The previous five years, practically my entire SAC career to that point, had consisted of lengthy Arc Light deployments alternating with nuclear alert at home. For the first time I settled into a "normal" crewdog routine, which was typically a five-week cycle with one week on alert, two weeks off, and three training flights in between. Alert was boring beyond belief, but was the core of the real crewdog way of life.

That routine was broken up a bit by satellite alert. SAC decided to spread the force out by stationing a handful of aircraft at outlying facilities, so from time to time we would pull alert with two bombers and two tankers at Mountain Home AFB, Idaho. That little fiefdom was ruled over with an iron fist by one Lt Col Hammond, who would assemble the crews in the briefing room and lecture them on leadership. He later gained some fame as the "Haircut Colonel" on Hawaii, who would intercept troops going to and from combat and detain them on the island until they got a proper haircut.

Alert, alert, alert! From 1964 to 1972, aside from Arc Light tours, sitting alert was my life. And then one day, 16 April 1972, I woke up on a Sunday morning in the middle of a typically dreary alert tour with a typically boring day ahead of me. I wouldn't know until years later that would be my last day on alert, ever.

The powers that be had decided Vietnam wasn't going well and they needed to get more B-52s into the fight. They stood down the alert force at Beale and other SAC bases and, under Operation Bullet Shot, deployed around 100 B-52 G-models to Southeast Asia to augment the D-models that had been carrying the load up until then. Beale was designated a Replacement Training Unit (RTU), responsible for transitioning crews from other versions into G-models. By then I was a highly experienced instructor and on a standboard crew giving check rides, so we stayed home. The new routine was fly, fly, fly, which I greatly welcomed.

I was able to perform one small service for the crews passing through our RTU. The B-52 Flight Manual consists of over 1,000 pages bound in loose-leaf binders, which was periodically updated with individual page changes which had to be manually posted. Each crew member had his own manual, had to keep it up-to-date, and would fail a check ride if his manual wasn't current. Crewdogs will recall, without joy, the many tedious page-count sessions it took to keep them out of trouble. I was the B-52 Manuals Control Officer, responsible for ordering and issuing these for the entire wing. Crews cross training from other versions had to get G-model manuals, which at the time consisted of a basic manual with about a dozen sets of changes, each change perhaps hundreds of pages in itself and often changing previous changes. The practice had been to hand out a stack of paper about a foot high and let the crewdog sort it out for himself. I thought this was incredibly inefficient and an insult to people going into combat. I got a couple of helpers, set up an assembly line, and put together complete manuals which I was able to hand out to pleasantly surprised crews.

Linebacker II, the intense bombing of Hanoi, kicked off on 18 December 1972 while I was still manning the RTU at Beale. On December 20th our squadron commander walked in, grabbed one of our younger officers, and told him, "We have some work to do." A Beale crew had been shot down with two fatalities and they were off to inform the families. Early Saturday morning, December 23, I got a phone call at home from the wing intelligence officer. It seems Brig Gen Edgar Harris, the 14 Air Division Commander at Beale, wanted to know why B-52s were getting hit and requested a briefing in his office. How was I supposed to know? I guess I got the call because I was the lead standboard EW at the time, but I knew nothing about how they were conducting the missions. Another Beale EW had been injured flying Linebacker a few days before and was already back home, so I very apologetically called him and asked if he could come along. He showed up in his Class A's, limping with a cane, and told the general about post-target turns and overall lack of tactics. I'm not sure the message got across, but it didn't make any difference anyhow, since the air division was essentially non-functional. All in all, it was a sad commentary on how things were done.

Incidentally, the nav who got shot down (and didn't make it) had until recently been on my standboard crew. We were getting our own check ride from CEVG, the SAC-level super-evaluators, and were about to make a routine entry into a low-level training route. This involved hitting a high-altitude entry point at a precise time, and as we were turning toward the point the nav's voice went up about an octave as he called out, "Tighten up the turn and give me all the speed you got!" He'd forgotten to allow for time in the turn itself, missed the entry, busted his check ride, and got fired from standboard. He was put on a regular crew, went Bullet Shot, and found himself in the wrong place at the wrong time over Hanoi. It was all for the want of a 4-minute turn.

In mid-1973 I heard from Dang, the U-Tapao Officers' Club cashier, who by then had saved up her earnings and was attending college in Scotland. She was about to go on vacation, had a summer job lined up in London, and had a new name: her Italian den mother in Glasgow called her Suki and it stuck. I myself was trying to get some leave time in Europe, so I wrangled a hop on a tanker as far as Mildenhall RAF, took a train to London, and after six years finally went out with Suki. It was okay with her, I guess, once she was out of her native habitat. Our first date was at Royal Albert Hall where we went to hear Stravinsky. She was planning to go back to Thailand, one thing led to another, and we ended up flying back to San Francisco together.

She stayed with me a couple of months at Beale and things were getting serious but we couldn't decide what to do about it. I eventually put her on a plane to Bangkok with the understanding that we'd wait a while and see what happened next. It didn't take long for me to see that was a mistake, and I went chasing after her.

Sixth Deployment

October 1973

Although I had a great job in standboard, I decided to give it up. By chance, as I headed out to put in my request, I came upon a Lt Col EW who was quite upset because his wife was going to leave him if he went TDY again. I said, "I think I'm going to solve your problem." I walked into the squadron commander's office and volunteered for Bullet Shot. There was nothing altruistic about it – I just wanted to get back to Thailand. Less than a week later I was on Guam, having traveled with a new crew via a commercial flight to March AFB, California, and a KC-135 tanker the rest of the way.

On October 25, five days after we arrived, we were in the SCAT school classroom when, at about two o'clock in the afternoon, a major walked in and told us to shut it down. Our forces had just gone into DEFCON 3 over some developments in the Yom Kipper War. Our contingency operations were over and everybody was being sent home. Fifty B-52s left the island over the next few days but we were not among them. Instead, on October 29, we boarded a chartered Northwest Orient 747 on the ramp at Andersen and taxied out. Most of the passengers were enlisted maintenance guys who had been living in the sweltering Tin City Quonset huts on base. A huge cheer erupted on board as we rotated and lifted off the runway. The Arc Light was finally out.

Epilogue

So what happened to the girl? I took some leave, got a hop, met up with Suki in Thailand, and we decided to get married. By fall of 1974 I was an instructor at Castle and we had our wedding in Merced, California. Forty years later we are still married and still there. At first the ladies of the Officers' Wives Club didn't quite know what to make of her, but within a year or so she was elected to the OWC board of directors. She went on to get degrees in nursing and in business, earned her certification in wound care, founded an Alzheimer's clinic and several wound clinics, and ended up as one of the most highly compensated RNs in the country. I have been around amazing people all my life, but this little girl who came from modest circumstances in northern Thailand is the most amazing one of all.

★ B-52s Fly 3,872 Combat Sorties ★

WESTOVER YANKEE FLYER

★ Tankers Mark Up 1,812 Missions ★

Published by the Westover Publishing Company, Inc., a private firm. Opinions expressed by the publishers and writers are their own and are not to be considered an official expression of the Department of the Air Force. The appearance of advertisements in this publication, including inserts, does not constitute an endorsement by the Department of the Air Force of the products or services advertised. The Westover Yankee Flyer is an unofficial newspaper, published weekly.

Vol. XV. No. 12. CHICOPEE, MASS. Friday, March 28, 1969

99TH RETURNS HOME

Flying with a Dysfunctional Arc Light Crew

Bob Davis

I had been stationed at Seymour-Johnson AFB, North Carolina, for about four years (1963-1967) when I received orders transferring me to Westover AFB, Massachusetts, home of the 99th BWG. My first shock of the move was finding out I was downgrading from the B-52G to the older D-model which had been modified with a larger bomb bay for the iron bomb mission. Previously the F-model had been employed for Arc Light but they all returned to the States in 1966. I was going to miss having my gunner partner sitting next to me in the new plane.

I was concerned because I left my position as a standardization evaluator and member of a select crew at Seymour to join up with a bunch of strangers and form a new crew at Westover. Luckily my pilot was also from Seymour and I knew he was a highly qualified one. Our co-pilot also came from there but, as I was to learn later, he was not so highly qualified. As for the rest of the crew, our radar navigator was an older lieutenant colonel who came from Castle where he had been an academics instructor and had not flown a meaningful mission in many months. The navigator was a passed-over captain from Hunter AFB who was lucky if he could find the crew bus. Somehow we came together sufficiently well enough to pass muster and to be called a combat crew before we were shuffled off to Andersen AFB, Guam.

The radar navigator was a real grouch who had not spent any time recently in a crew environment (meaning he had not been on B-52 alert in a long time.) Not only was it difficult for him to place the cross hairs on the target while flying, but even more difficult to become an integral member of the team. I will never forget the night we short-sheeted his bed and rigged it so the legs would collapse. I don't remember the circumstances now but we knew he would be the last one to come to bed. What an eruption occurred when he tried to slide under those covers, hit the bottom and banged around just right so the legs collapsed. He threatened that a court martial would be our just reward. But let me go on to more meaningful exploits. I'm only going to recount a couple here since they will be sufficient to demonstrate how lucky we were to come home from our tour in one piece.

Me on crew bus

The first departure from the normal activities came when we were flying as number three in a three-ship cell. That was our usual position because of the skills of our RN/Nav team, and on that flight we were scheduled to do a timing bomb release off the first aircraft. I'll admit to not remembering the different types of bombing missions we flew, but that was the occasion when the number one aircraft acquired the target and the other two ships dropped their bombs off of the lead aircraft's timing. Well our Nav team got the timing down but they measured off of the number two aircraft instead of the number one. Since we were on the border of Laos and Vietnam, we ended up dropping our bombs in the wrong country. I believed that minor indiscretion was later attributed to some strange mortar attack.

The second episode prompting me to seek reassignment to another crew came as we were returning back to base after another mission. We were on final approach into U-Tapao RTNAF, Thailand, when the pilot asked for flaps. Well, to our co-pilot one lever was as good as the next so he deployed the drag chute instead of lowering the flaps. Of course the chute was sheared from the aircraft and we had to land without the benefit of a drag chute. Following the de-briefing, the co-pilot was assigned a vehicle and told to back track and try to locate the chute but as I recall he was not successful. By then our crew was already gaining notoriety, the type of which I was not really too happy to have. I informed my AC (a good friend and fellow Virginia native) I was going to ask for a transfer to another crew.

He asked me to please not leave him since the gunner and I were the only ones on the crew he could rely on. Well I stayed and we managed to make it through the rest of the tour without any further incidents. We even managed to fly one mission up North with some hairy North Vietnamese SAM activity in the area that earned us a DFC. We eventually made it back to Westover in one piece and I managed to get an assignment out of BUFFs and into some really interesting reconnaissance work.

Needless to say, I will never forget the Arc Light tour with those misfits from the 99th BWG.

Me (left) being sworn into the Regular AF from Reserve status by my A/C Harry Lee.

EB-66

The SAC Guy

Pat Branch

Grandpa, what did you do in the Vietnam War?

Recently my granddaughter's 5th grade teacher asked her class if they knew any veterans. When asking, my granddaughter was surprised to learn that I, her own grandfather, was a combat veteran of the Vietnam War. I'm glad they ask questions such as this in school. I think it is important for our children to know that their country has been formed by the commitment of American Veterans, including members of their own families.

During our discussion, one question seemed to lead to the next. A logical follow-on question during our talk about my military awards, and one I have pondered several times myself, is what I did to deserve earning the Distinguished Flying Cross (DFC).

Before I go any further, let me assure you I know this book is supposed to be about B-52 exploits, and I want to make it clear this story would not be possible had it not been for the time I flew in the BUFF and is very much concerned with B-52 operations in Southeast Asia. The photo above is not a B-52, of course; it is an EB-66E.

I had rotated back to the states and had been newly assigned to the RC-135 program at Offutt AFB, Nebraska. No one was more surprised than me when, during a 343 SRS Commanders Call, the CO announced the names of those selected for promotion to major and presented medals to newly arrived crew members. I was a bit relieved, but fully expected to be on the major's list and the two Air Medals awarded were common place in the 55 SRW. The presentation of the DFC, however, was a complete surprise. I was hoping to receive the Bronze Star. I knew the Chief Master Sergeant of the ECM Maintenance Squadron had submitted me for the Bronze Star. Bronze Stars were generally reserved for non-flyers as end of tour medals for those who served with exemplary accomplishments while in the war zone. The Chief and I, acting as the Wing EW, had worked several projects together. I was impressed and flattered that he would go out of his way on my behalf. On the other hand the DFC was normally awarded to truly heroic combat pilots, which I was not.

The DFC story begins with my first combat sortie in the EB-66E. By 1972 the EB-66E/C aircraft were old airplanes; supply sources no longer existed. A few B-66 were still at Shaw, their stateside training base. The rest of the fleet had been deployed to Korat AB, Thailand, to be used as spare parts. Due to a lack of aircraft, the training of EWs at Shaw was simulator-based and the crews received no flights in the airplane while there. My first ride in an EB-66E/C was a combat sortie in late October, 1972. Both the Nav and I had been recently assigned to the unit. The pilot was a Lt Col on his 3rd combat tour.

By the time of that first EB-66 flight I was a Captain with eight years of service and seven years as a rated navigator and Electronic Warfare Officer (EW). I had about 2,000 hours in the B-52 and had flow B-52B, D, F, G, and H models. I had also logged 350 hours in the B-58. I had one B-52D Arc Light tour of 104 combat missions and had just spent the previous 18 months as the T-4 (EW Simulator) operator/instructor at Grand Forks AFB. In that position, my job was to ensure Grand Forks based EWs were ready for their Arc Light tours. In short, I knew the B-52 missions and their tactics. As it turned out, I knew them better than anyone else stationed at Korat.

The mission of the 388TFW, Korat AB, was to fly the Tiny Tim package supporting the B-52 missions. Tiny Tim was code word for the F-4 MiG Cap, F-105 and F-4 Wild Weasels, and EB-66 ECM aircraft.

So, off we go into the wild blue yonder – an experience pilot, a new Nav, and new EW. We crossed over Vietnam to the Gulf of Tonkin, and set up an orbit to parallel the track of the incoming B-52s cells targeting Vinh. I set up the EB-66E's 23 jammers according to a pre-set checklist used by all the EWs in the squadron - a checklist I was about to rewrite in the coming weeks. As the enemy radars came up, I centered each jammer just as I would have done had I been flying in the lead B-52. The BUFFs checked in with ABCCC. Very shortly thereafter the first SAM signals appeared on my ALR-20 receiver. I refined my jammers and notified the pilot.

The conversation when something like this:

"Pilot, EW. SAMs at 12 O'Clock."

Pilot's response – "Good- it will be a short night. BUFFs don't fly where there are SAMs."

A few minutes later:

"Pilot, EW. SAMs up again, strong signal 11:30."

Pilot – "Crew we are headed home. The BUFFs will be turning back."

EW – "No Sir; the BUFFs are IP inbound. They will hit the target."

Pilot - "The BUFFs must be late and not at the IP."

EW – "No Sir, the BUFFs are well past the IP; they are on a 'Press-On'sortie. They will hit the target - missiles in the air or not."

There is a long pause before the pilot's response.

"Crow - how many 66 sorties do you have?" He knew the answer of course.

With a chuckle I responded "This is my first sir."

Pilot – "Well this is my third tour over here. BUFFs don't fly with SAMs."

"Sir, they have a special category mission called a press-on sortie. If they were going to turn back they would have already done so."

The rest of the flight was in silence. The pilot said no more, and I didn't dare push the subject. When we landed the pilot went off with the Supervisor of Flying, and the Nav and I when to debriefing.

After the debriefing, I was walking back to the Weapons and Tactics office to visit my friend Dick Sherman, when I overheard my pilot talking to Col Chico (first name), the 388 DO. He was saying, "That new Crow knows what the BUFFs are going to do before they even do it." About then, the Colonel spotted me walking by.

"Branch, get in here!"

That was my first meeting with the DO, and I reported in using the traditional military format, which he wanted none of. I briefed him on my B-52 background and told him how press-on sorties were covered by the SAC tactical doctrine. A week later, I had a new job. I would remain assigned to the 42TEWS as a crew member but with an additional duty assignment in the 388 TFW Weapons and Tactics office replacing my friend and former instructor Dick Sherman. Dick had been working on a revised and more effective standardized jammer settings list. I had followed his draft setting on that first flight and would later publish it for all EWs to follow before the beginning of Linebacker II in December of 1972.

That first EB-66E mission shaped my year at Korat and ultimately impacted the development of my Air Force career, but that is a much longer story. Some 43 years later, I find I have difficulty remembering dates and names from then. In November or very early December of 1972 a briefing team was set up to convey "Fighter Pilot" wisdom and experience to SAC in-theater leadership. The driving concept was that experienced crews flying MiG cap and Wild Weasels were concerned SAC tactics used to bomb the Ho Chi Minh trail would result in heavy losses if employed against targets in the Hanoi or the Haiphong harbor area. The briefing team consisted of Jack Lippold, 388 Weapon and Tactics lead, Bear Gleason F-105 EWO, and Jeff Feinstein, F-4E ACE.

While I was not a briefer, I was selected as part of the team because I was "The SAC guy." All the aforementioned officers were also Captains. We did have adult leadership. Here memory fails me, but I am rather certain that among the accompanying Colonels were the 388 TFW/DO and the Wild Weasel Squadron Commander. Both were extremely experienced and highly effective combat pilots. Their mere presence, along with Feinstein recently becoming an ACE, gave the team creditability. We took the briefing to the B-52 Provisional Wings at U-T and Guam. Before taking the briefing to SAC, we first briefed our own Wing CO, the 7/13 CO, (NKP), and 13 AF/DO (Clark).

I particularly remember the briefing we presented on Guam. The Division Commander and his full staff received the briefing. A very self-assured (cocky) Feinstein told the General that if SAC did not change their tactics they would suffer horrendous losses. The briefing recommended hitting multiple targets at the same time and avoiding the big turn right after the target. As we all know that message fell on deaf ears. I don't know who changed those tactics or how SAC made the decision to radically change their tactics midway through Linebacker II, but I do know it wasn't the first time they heard that message. From the briefing at U-T we returned with copies of the SAC tactics documents. I wrapped them, signed for the documents, and carried them back to Korat all in accordance with the existing instructions on handling classified material.

Jump forward in this story to 22 November 1972 and the EB-66E ECM support to a cell of B-52s bombing just west of Vinh. The Nav and I had planned a route optimized to cover the BUFF's targets and briefed the pilot (Aircraft Commander), who approved the track and orbit. The B-52s hit their target and began to checkout with ABCCC.

"Lead off target RTB."

"Two off target RTB."

And then nothing.

The ABCC and Lead both repeatedly called three without an answer. The deafening sound of that vacant call is extremely haunting. I would hear that sound of emptiness again during Linebacker II and later while flying in an RC-135 out of Alaska when our tanker bought the farm shortly after take-off after sitting on the ramp for four hours in minus 54F temperatures while waiting for our RC-135 to stop leaking fluids and take off on a peace time strategic reconnaissance mission. But I digress - back to Vinh.

This was the first loss of a B-52 to North Vietnam SAM II missiles. We later learned the BUFF crew had made it back over Thailand and bailed out near the US Base at NKP. The Tiny Tim support packaged included two Wild Weasel aircraft and an EB-66E sortie. When it became obvious the B-52 was shot down, our pilot ordered the Nav and me to replan our orbit and cut the distance between us and the BUFF targets in half. After responding with a hardy "Yes Sir," I reminded the pilot the new orbit would put us in the middle of the active SAM lethal range. The pilot's response was a simple "I know." The two Wild Weasel crews picked up similar orbits and we held our orbit until the second Wild Weasel crew declared emergency fuel and RTB'd back to Korat. That left us alone and unarmed in the middle of an active SAM ring - it was time to get out of there.

My story now jumps forward to Linebacker II in December of 1972. Being assigned to the Weapons and Tactics office, we had 24 hours advance notice of Linebacker II plans. One of my main tasks that day was to locate as much chaff as we could find on base and get it up to the RF-4 photo reconnaissance squadron flying out of Udorn AB, Thailand. The RF-4 was tasked to lay a chaff corridor to mask the route of the penetrating B-52s. The maintenance chief and I literally crawled through ever shortage bin on base, and by the time the crews were ready to fly, I had worked a 24-hour duty day. After the first wave took off, I went to my hooch to get a few hours of sleep and would be back four hours later for debriefing. I was not surprised that I was not on the flying schedule those first two days of Linebacker II, but when my name was not on the schedule the third day, I when to the scheduler to find out why. The Squadron Commander had directed that no one with less than three months in the B-66 would be allowed fly a Linebacker mission. I suppose it made good sense to him, considering the abbreviated training program used for EWs. Still, it made no sense to me, so I went to see the commander.

"Sir, I understand that it is your orders that no one with less than three months on station is to fly Linebacker."

"That is right Pat. Do you have a problem with that?"

"Sir, I don't want to fly those missions any more that the other guys. But this doesn't make any sense. You are sending new lieutenants up to fly just because they have been here more than three months and leaving more experienced crew members on the ground."

I reminded him of my B-52 Arc Light time and that the BUFF crews flying were my buddies. I flew five out of the remaining eight days of Linebacker II, becoming the high time flyer among the EWs for both Linebacker II and for the year I was in the 42TEWS. The Air Force may deny that a DFC was ever awarded as an end-of-tour medal, but that is probably the day I won mine.

Jump forward to 23 Dec 1972. When I came into the mission planning room for my second of five Linebacker II missions, I was informed by the scheduling officer I would be the reconnaissance team lead on a EB-66C ELINT mission. Our task was to find a signal know as the T-8209. It was a modified Fire Can Radar operating in the India Band frequency and was suspected of providing the North Vietnam SAM sites with accurate elevation data. The North Vietnamese also sent MiG aircraft up to collect altitude information as well. The T-8209 had been recorded by Wild Weasel crews during Linebacker I, almost a year earlier. In fact, I played that recording for all the newly assigned 388 TFW crews as part of their "new-guy" orientation. The PACAF Electronic Intelligence (ELINT) Center later concluded the signal never existed. They were and still are wrong.

I was one of only a few people in the squadron who had qualified as a reconnaissance crew lead and was the only one available on that date. I had an advantage, the EB66 C-model I was assigned had very old, out-dated equipment. It was the same equipment I had flow with on the earlier model B-52s. We had the APS-54 RHAW, ALR-9 and ALR-14 receivers. To get a satisfactory score on a SAC ORI, EW's had to manually tune both receivers at the same time - listening to the ALR-9 while watching the ALR-14.

Why was that important? On 23 December 1972, the EB-66E I was originally schedule to fly crashed and burned while returning to Korat AB.

We landed in the EB-66C and walked into the Squadron Ready Room without knowing of the loss. The lieutenant on duly had just written all of our names on the scheduling board as "Missing in Action." The E-model had been the more reliable airplane. The C-model had very poor self protection ECM and a habit of not always coming back. Bat-21, a C-model, had been lost the previous April, setting up the longest search and rescue operation during the Vietnam War. Of the seven crew members aboard, only the Nav survived.

23 December 1972 is a day that has lived in my memory for the past 43 years, partially because we lost that crew, but also because I felt an unwarranted sense of guilt. I lived that day and have enjoyed a very good life - because, someone else did not. By the way, the official record of that flight and the "Wall" in Washington, D.C. are wrong. Hank was the Nav, not the EW. Bill was the EW, not the Nav.

After the Christmas stand down the B-52 changed their tactics. Rather than stream in one cell after the other on the same heading and altitude, they hit their targets from two different directions at the same time. While this raised the probability of a mid-air coalition (a SAC fear based on the first B-52 Vietnam raid) it gave them a much shorter exposure to the SAMs. It was on one of those flights in the EB-66E that we encounter a true hero of the air war, our assigned tanker.

Again we had trouble getting off the ground on time. Our mission was over the Gulf of Tonkin, near the city and harbor of Haiphong. To make up time we slipped between the SAM rings near Vinh to refuel slightly north of the Vietnam DMZ. Unlike some other aircraft, the EB-66 used a probe and basket refueling system. There was only one KC-135 tanker available so equipped. We met our tanker over the Gulf and started to take on fuel, but reached the end of the refueling route before taking on all we required. The EB-66E pilot asked the KC-135 pilot to continue north until he could get all of the gas. No Way! The KC-135 Pilot stated that he was under strict orders not to go north of designated latitude and we were there. We took on the remainder of the gas in the turn, dropped off the tanker and continued on our way.

That was a busy night for the BUFFs hitting targets in the Hanoi area. We were positioned just off the coast below their ingress and egress routes. We were using a large island as our turn around point. According to intelligence there were no active SAM sites on that island; however each time we approached the island, I took a "look-through", by quickly turning off selected jammers to be sure I knew what signals those jammers might be masking. On our second or third orbit my precautionary look-through revealed the strongest BG-06 (SAM Guidance) signal I had ever seen. We were obviously the target of a SAM site that was not supposed to be there. We turned short and immediately adjusted our turn-around point.

Perhaps the most memorial event of that night was the reaction of the B-52s as they exited North Vietnam. Thanks to improved tactics and the shortage of North Vietnam SAM missiles, no B-52s were shot down that night. Coming off their targets but not yet out of harm's way, as the BUFFs passed over our position their EWs picked up our navigation radar and assumed we were a North Vietnam MIG. To defend my fellow EWs, the B-66 nav radar did sound a lot like a MiG radar in the search mode. The BUFF EWs unloaded their chaff and flare dispensers over us that night.

At the end of the mission we declared RTB and exited our orbit. Almost immediately our tanker was on track at the southern end of our orbit. Our Pilot plugged into the tanker's basket, started taking on the gas and called the KC-135 pilot.

"Hey aren't you the same son-of-a-bitch that wouldn't go even a few miles past the end of the refueling anchor?"

“Yes Sir.”

“Well what are you doing up here.”

The KC-135 pilot responded, “We have been listening to the radios and thought that you were in trouble."

To me the true hero has never been the smartest, bravest, or most skilled warrior. It has always been the warriors who have extended themselves beyond their normal capability or that which can reasonably be expected of them.

Our final engagement with SAM missiles came several weeks after Linebacker II. The B-52’s target was a supply depot just north of the DMZ. It was in the same area where Bat-21 had been shot down in April of 1972. The nav and I had planned a route to cross under the B-52’s track just outside the known SAM ring. We presented the track to our Aircraft Commander expecting approval. Having been in the Squadron when Bat-21 was shot down, the pilot immediately rejected our plan. At his direction, we moved the orbit farther to the west - well outside any known SAM ring but still crossing the B-52’s track. Our objective was always to look like a full B-52 cell of three aircraft to the North Vietnam radar operators.

When we flew the mission and reached the point where our track and the BUFF track intersected, the pilot (the only one with a clear view outside of the aircraft) came over the intercom.

"Damn you EW - here they come."

"Here what comes Pilot?"

"SAMs - three of them."

The first SAM went high toward the BUFFs. The second missile was a dud and failed to gain altitude. The third SAM came toward us and at our altitude. By that time of the war the pilots believed if they could see the missile, they could out fly it. Rather than doing a textbook SAM maneuver our pilot kept moving our aircraft away from the trajectory of the missile. If we had flown the track I had planned, we would have been easy prey for the SAMs coming out of the same site that destroyed Bat-21, but which had been inactive for almost a full year.

An honest analysis of this engagement would conclude that the SAM missiles were targeted on the BUFFs. The first missile was directed at the altitude the BUFFs were flying and the second missile aimed at where they expected the BUFFs to be after an evasive maneuver. Our B-66 just happened to be there. It was never the intended target.

Official USA Photo

First Bomb Run

Jay Lacklen

When asked if I killed anyone during the Vietnam War, I always have to answer, “I don’t know. I *may* have, and probably did, since my B-52 bomber crew dropped several thousand pounds of bombs on the Cambodian jungle.” I will never actually know.

I only flew two live-bombing runs in March of 1973 before President Nixon’s final bombing halt prior to the negotiations ending our involvement in the war a few months later. But the first mission I flew captured all the terror, anticipation, wonder, and angst of flying into a real war.

The mission began after dark at Andersen AFB, Guam, in the western Pacific Ocean. The crew bus dropped us at the hot loading zone on a far corner of a field. It was located in a remote site to provide some protection for the base if something went wrong while loading over 50,000 pounds of bombs in the bomb bay and on wing pylons of our B-52D.

A full moon painted the bristling black war bird in a ghostly light - a formidable metal dragon that would righteously drop explosive mayhem onto America's enemies, or so I thought at the time. I paused and slowly put down my flight bag and stared in awe. Was I really a part of this? What was I about to do?

Six hours after takeoff, we approached the target area over the Parrot's Beak region of Cambodia as one of a half-dozen three-ship-cell bomber formations. Each cell was named for a tree. We were Oak Cell; others were Pine, Maple, and Birch.

As the lead aircraft co-pilot for my cell (for some unknown reason, I got to be lead on my first mission), I had to announce the impending bomb drop on "Guard," the international radio frequency all aircraft monitor. The call allows aircraft in the vicinity to vacate the area and avoid the "rain" of our falling bombs. (As an aside, this is the same rain referenced in the Credence Clearwater Revival song "*Have You Ever Seen the Rain?*")

I had been cautioned to switch my radio toggle from the interplane frequency to the Guard channel before transmitting the warning. Being a raw rookie, however, I gave the entire two-minute spiel on interplane, to the great amusement of the other co-pilots. "Hey, lead, want to try that on Guard?" they snickered on the interplane frequency.

Then the bomb run began in the early morning darkness over an Asian jungle. Our three-abreast, triangular-shaped formation banked steeply, ominously, onto the bomb run heading. The radar navigator, who would throw the switch to drop the bombs, informed the crew we were approaching the Initial Point (IP) to begin the run.

Just as we passed the IP, a male Asian voice began transmitting in Cambodian on our radios. He sounded as if he was babbling in an opium den, and his voice disturbed and frightened me. I feared he might be an apparition warning us off our task - a voice of doom making his last hopeless statement to his slayers to offer a last opportunity to save ourselves, or a soon-to-be victim in the target zone. He continued talking, as if relating a story to a fellow opium smoker, while our formation approached the target. No matter what I did to my radio controls, I couldn't make the voice stop.

In the near distance, 33,000 feet below us, the ground below glowed an eerie red from the explosions created by the preceding bomber formations. A mist hung over the terrain giving the area the look of a graveyard in a horror movie.

The radar navigator began the countdown to our own bomb release, "Ten . . . nine . . . eight,"—the apparitional voice continued his drunken soliloquy on the radio—"Three . . . two . . . one . . . bombs away!"

The aircraft shuddered lightly as the bombs unhooked from the wings and dropped from the internal bomb bays. Our three aircrafts' ten-second release sequences could obliterate an area equal to three football fields and unleash a shock wave that would kill any unshielded creature within half a mile. North Vietnamese soldier and author, Bao Ninh, later wrote the immediate aftermath of such a strike resulted in "a rain of arms and legs dropping before him on the grass."

We waited as the radar navigator counted down to detonation, about 50 seconds for the bombs to fall the 33,000 feet. "Three . . . two . . . one . . . impact!" The thin clouds around our aircraft reflected hundreds of small bursts of light from below. It was done.

The radar navigator announced the closing of the bomb bay doors.

We flew on in the darkness in silence as I pondered the damage we had done.

I ponder it still.

From "*Flying the Line, an Air Force Pilot's Journey*."

BUFFs Over Korea
Mike Brinkman

After having been a spare nav for a year and a half at Robins AFB, Georgia, I was looking forward to finally being assigned to a crew and my wife, Kathy, and I were both excited about getting a permanent change of station (PCS) to the Western Pacific.

My new assignment was to the 60th Bomb Squadron, 43rd Bomb Wing, at Andersen AFB, Guam, where I served from May of 1976 until May of 1981 as a B-52D ("Dog" model) navigator and instructor nav.

Our arrival on Guam, however, was less than auspicious. As we finally exited the Pan Am 707 "cattle car" at 0600 local, two things got our attention: (1) The slap in the face from the heat and humidity, and (2) the greeting from a pair of smiling ladies: "Welcome to Guam; here is your typhoon preparedness pamphlet!" What?

Upon entering the terminal, we were greeted by our sponsor and his wife. He informed us he had to leave because the wing was evacuating all the aircraft to Kadena AB, Okinawa, as a precaution due to an approaching typhoon. But we were not to worry, they had already had two false alarms that season (typhoons that missed), and he should be back in a couple of days. In the meantime, we were invited to stay with his wife on base. Needless to say, the slow-moving Super Typhoon Pamela did not miss the base, but ran right over it, chewed it up, and spat it out instead. Not only was Pamela's slow passage over Guam terrifying (though the passage of the eye was sort of cool), but afterwards the island spent four miserable days without electricity and running water. Welcome to Guam indeed! Well, that is a story for another time.

Fast-forward a few months to 18 August 1976. That was the day I was due to take my D-model checkride. On the way into the base from DOD housing in South Finegayan, I heard a radio news report that two U.S. Army officers had been attacked by North Korean soldiers on the DMZ and had both had died as a result of their attack. I considered the shocking news for just a minute, but quite frankly, the checkride was the main thing on my mind.

When our crew arrived at the airplane that morning, the pilot discovered it had too much fuel for our planned mission. While the excess fuel was being offloaded, we crewdogs sat in the back of the maintenance line chief's "bread truck" (step van) talking about the now necessary changes to the flight plan (more stress on me) along with the news from Korea. About then, the line chief's maintenance net radio squawked, asking the chief if his crew was with him. After an affirmative answer, we were told our sortie was cancelled and we were instructed to grab our gear and report to the squadron briefing room immediately.

Most of us had a pretty good idea what all the hub-bub was about. The briefing room was packed – it must have given the old heads flashbacks to Linebacker II days.

We were called to attention as the commanding general of the 3rd Air Division took the platform followed by our wing and squadron commanders. An officer from the intelligence shop then proceeded to give us the lowdown on the situation in Korea. Two U.S. Army officers, Capt. Arthur Bonifas and 1st Lt. Mark Barrett, along with their Korean (ROK) counterpart, had supervised a crew of 11 American and Korean enlisted personnel in the trimming of a large poplar tree. The tree was located near a United Nations Command (UNC) checkpoint guard house. The checkpoint was located in the Joint Security Area of the DMZ at the northernmost end. A bridge, half of which was in the South and the other half in the North, was known as the "Bridge of No Return." The south side of the bridge was where the checkpoint was located and was a dangerous place to guard. The only way to keep an eye on the checkpoint and its guards was from an UNC observation post up the hill from the bridge. The foliage from the tree blocked the line of sight between the observation post and checkpoint, so there was a need to prune the tree.

On the morning of 18 August 1976, the tree trimming crew began their work. A North Korean officer with about 15 soldiers crossed the bridge and, after observing for a while, the officer objected strongly to the pruning. When Capt. Bonifas ignored his protest, the North Korean officer sent a runner across the bridge to return with a truckload of reinforcements. Using the tree trimming crew's own axes, the North Koreans attacked, fatally wounding both U.S. Army officers and injuring all but one of the UNC soldiers.

After the intel officer finished with his briefing, the general addressed the crews, followed by the squadron commander. The gist of the briefing was Washington decided this blatant provocation needed a strong show of force in response, the plans for which were being prepared. All the flight-worthy bombers and tankers were to be generated (prepared for flight), and I was assigned to a generation crew. We would then wait for orders to launch with or without bombs and ammo.

I remember sitting in the nav seat on the BUFF we were generating about 0200, when, out of the corner of my eye, I saw a flash of blue heading up the ladder to the upper deck. A minute later, the general stuck his head between the radar navigator's and my seat, looked us each in the eye and said, "Are you boys ready to carry iron?" We answered in the affirmative, of course, though we later agreed that carrying iron (conventional bombs) to targets in North Korea was not something we were eager to do.

Within a couple of days, Operation Paul Bunyan was underway and we were flying three-ship cells of B-52s from Andersen up to the DMZ; the first BUFFs to fly over Korea. However, we weren't alone. In addition to us and the USAF F-4 Phantoms and Republic of Korea Air Force (ROKAF) F-5s and F-86s, there were F-111 fighter-bombers from stateside and F-4's from Kadena AB, Okinawa, and Clark AB, Philippines, joining the mix. To add frosting to the cake, the aircraft carrier USS Midway task force was also stationed off shore.

With an instructor navigator onboard, I was allowed to fly on one of the early sorties. The good news is that we were not "carrying iron." The bad news is that we didn't have any ammo for our tail guns either. So, except for our electronic countermeasures, we were defenseless. Fortunately, we had an excellent MiG CAP of fighters protecting us from North Korean fighters; however, the North Korean forces did try to electronically sucker us into flying north of the DMZ with false navigational signals. That was not our main concern. The weather over the peninsula was crappy to say the least. With our three-ship cells flying parallel to the DMZ and all the other aircraft in the air, we were more concerned about the effectiveness of command and control, and we were aware of the potential for a midair collision. To say that the crews were stressed is an understatement. Thankfully, nothing bad came to pass, and all bombers returned home safely.

Operation Paul Bunyan was a success. In fact, a group of well-protected UNC engineers returned to the subject poplar tree and reduced it to a 20-foot stump while a well-armed group of North Korean troops watched. The stump was cut down in 1987 and a monument in memory of the slain Army officers was erected on the site.

Interestingly enough, after the crisis was over, the U.S. State Department requested that the 3rd Air Division fly weekly B-52 two-ship cells from Andersen to Korea as a continuing show of force. The Pentagon complied and the BUFFs flew on those missions beyond my tour on Guam. Eventually the bomb wing took advantage of this opportunity to maximize training. The 43rd developed a low level route that ran the length of the peninsula thus allowing us to fly over mountainous terrain instead of just over water. A bomb scoring site was put in place so that we could do electronically-scored bomb runs.

The sortie profile for this 12-hour mission included all the standard elements: a celestial navigation leg, air refueling, low level navigation, and bomb runs. So, what was the difference?

1) We were flying in the vicinity of real anti-aircraft sites and initially the Electronic Warfare Officer (EWO) was allowed to jam the sites. After awhile though, the Army had us quit the jamming due to the effectiveness of the BUFF's systems.

2) Poor navigation heading north could put the bomber in the crosshairs of North Korean fighters or SAM sites. There was more than one sortie where less than sterling navigation resulted in North Korea scrambling fighters.

3) After the bomb runs, we would climb out to altitude over the China Sea to the west of the Korean peninsula. Running on an east-west track, we provided an intercept target for USAF and ROKAF fighter aircraft. The fighter jocks loved it. We didn't so much. Why? First of all, at that time, all fighters in the air were armed with missiles and guns. Also, there was a 150 knot jetstream blowing from west to east. The Nav had to really pay attention while flying into the head wind. On one flight, I wasn't paying proper attention only to look up and see the coast of China on the radar. The RN and I decided that it would be a good time to turn back east.

Finally, just to keep things interesting, on several occasions we diverted into Okinawa on our way home from Korea due to a typhoon approaching Guam. A couple of times we did this only to discover that the rest of the B-52 fleet didn't launch. Imagine how disappointed the Okinawan anti-nuke demonstrators at the approach end of the runway were when no other bombers showed. The demonstrators always seemed to know when we were coming. All in all, it kept life in the Dog model interesting.

Three years after The Tree Cutting Incident, I was among the group of crewdogs who were the guests of the ROKAF. We were flown up to the DMZ in an Army CH-47 Chinook helicopter. Initially, we landed on a postage stamp size LZ on the side of a mountain and climbed up to an observation post to look over into the North. We were then flown to Camp Kittyhawk (later renamed Camp Bonifas in honor of the slain captain). There we were briefed before being bused to a walking tour of the Joint Security Area. The entire time our group was under the watchful eyes of severe-looking North Korean guards. I have photos of them taking photos of us. It was spooky. They truly despise us. We were then transported to the observation post overlooking the site of The Tree Cutting Incident. The necessity of the tree trimming became clear from that vantage point. But what was most significant to me was when we rode down the knoll and around the stump of the tree itself. To this day I regard that site as holy ground.

Photo by my RN, "Boonie" Bill Meyer of me walking to my B-52D to lead the last Arc Light cell out of U-Tapao AB, Thailand in June 1975

Putting Out the Arc Light

Jay Lacklen

At the end of May of 1973, the B-52 Arc Light operation ended as the final three-ship cell, with our crew in the lead, departed U-Tapao RTAFB, Thailand. For many years, SAC crew members had joked about "putting out the Arc Light for good," and finally we were about to do just that. I had only flown two actual Arc Light bombing missions in 1973, and the rest of the crew had flown none. Still we had the honor to put out the "Light" for the thousands of crewmen who had flown the tens of thousands of hours over the previous ten years. They had done the heavy lifting; we had done little for the time we were at U-Tapao, but we got to take it home for good.

We would fly in a three-ship cell formation until we reached the coast of California, then "two" and "three" would leave the formation to land at March AFB, California, and we would continue on as a single ship to Carswell AFB, Texas, a 16-hour mission.

We had an indignity to suffer before takeoff, however. The staff told us to report to our aircraft an hour early so we could string up a room full of rattan furniture in the bomb bay to deliver to the Deputy Commander for Operations (DCO) at Carswell. That did not sit well with my crew. We already faced a tortuously long mission without any extra duties, and we saw that as an absolutely unnecessary task. No doubt one of the commanders at U-Tapao garnered a significant kudo for shipping the colonel's furniture for free, but we were the ones providing the service. We hoped we'd have to cycle the bomb bay doors en-route over the Pacific - one final rattan bomb run.

When the time came, the three B-52s sat behind the runway hold line awaiting permission for the final B-52 takeoff from U-Tapao. Then tower cleared us for takeoff. On interplane, I said, "OK, guys, let's take our BUFFs and go home," and with that statement we took off. As we climbed out over the Gulf of Siam, U-Tapao fell silent for the first time in a decade and would largely remain that way in the future.

At 42,000-feet cruise altitude over the Philippine Sea, two interesting things happened. First, a jet contrail passed over us. That was unusual because we were already at our max altitude of over eight miles high, and the other plane's contrail was so far above us I couldn't see the aircraft. I decided it must be a U-2 or SR-71 spy plane because I didn't expect to encounter any aircraft higher than we were, much less so far above us I would be unable to see it.

Next, a US Navy ship broadcast on guard channel asking if any American military aircraft was in the area. We answered that we were, and he asked if we could go "cipher" mode - the scrambled, secret mode on one of our radios. The ship said they had an unknown target apparently shadowing them and asked if we could drop down and take a look. I said I'd have to get permission to do so because I couldn't make my destination if I used the extra gas to chase a shadow. Shortly thereafter, the ship reported they had resolved the issue and we wouldn't be needed. I thought, briefly, we might find an excuse to cycle the bomb bay doors to rain rattan on the mysterious shadow, but it was not to be. The Carswell colonel would get his damn furniture.

As the final indignity, neither colonel (at either end) met us to thank us for our efforts.

From *"Flying the Line, an Air Force Pilot's Journey,"* by Jay Lacklen.

Chapter Five– Southwest Asia

Southwest Asia - [south-west] [ey-zhuh] – *noun* - the part of Asia that includes the Middle East.

Official USAF Photo

The Butchers of Mosul

Chris Buckley

Disclaimer: This event occurred many moons ago. The story has been told several million times over many more beers. Please excuse me if I have misquoted anyone or mixed up facts.

There I was - 39,000 feet over Bagdad.

I was the Navigator for Crew-12; collectively known as the Butchers of Mosul. We got that name because on every sortie we flew into Iraq we found some reason (however ridiculous) to drop at least one bomb in or around the city of Mosul. Every one of the 20 bomber crews from the 23rd Expeditionary Bomb Squadron had some kind of moniker: The Regulators, the Gunslingers - you get the point. We were also known as the JV crew, as in Junior Varsity because we were not exactly the best on the line, if you get my meaning.

It was early April of 2003 and we were on our fourth combat sortie in Operation Iraqi Freedom. Our mission was to fly B-52H tail number 60-0060 "Iron Butterfly" from RAF Fairford to Bagdad and drop twelve 2,000-pound satellite-guided bombs. Each bomb would be striking its own target, allowing us to destroy 12 different things at the same time.

We had just air refueled and were very heavy, around 400,000 pounds gross weight. We droned easily towards Bagdad and saw several contrails from surface to air missiles. That was nothing new, since the Iraqis had been lighting off SAMs for some time. They were all unguided since the Iraqi forces were deathly afraid to radiate anything with a guidance radar lest it shortly met an anti-radiation missile. So seeing missiles flying was both scary and not really that scary at the same time. Our Electronic Warfare Officer (E-Dub) was the twitchy sort, excitable and nervous who stayed fully strapped into his ejection seat, always wearing his helmet and survival vest. The Aircraft Commander, our Pilot, was not excitable, nor was he nervous, nor was he strapped into his seat. He flew in "cruise comfort," meaning his ejection seat was in the safe position and he had unstrapped from his parachute. He wore his headset instead of helmet and his survival vest lay tossed across the crew bunk. The aircraft was on autopilot, the pilots barely taking notice of the outside war. That was when we all heard it. WHOOSH! This sound was loud enough to be heard through the helmet, the ear plugs, and over the 100 decibels of engine noise that continuously filled the cockpit.

"What the hell was that?" The E-Dub asked.

"SAM," the Pilot remarked casually. "Off the left side of the nose."

"Holy Shit!" The E-Dub called out. "Break Right!"

"Why? It already missed us."

At that point I didn't know whether to laugh or cry. But the bombs were ready to fly and Bagdad was getting closer, so those who were not already strapped in got there. We approached our drop zone and the computer told us the bombs were safe and in range of the targets. The Radar Navigator gave me a nod and I flipped up the plastic protective guard on the button marked MANUAL LAUNCH and pressed the button. You could feel the airplane shudder with each weapon release. In five seconds we had dropped 11 of the 12 bombs.

"Looks like we got a hanger," the Radar Navigator called to the crew as the co-pilot put the plane into a left turn.

Like I said, we were heavy and we were high. So when the left turn hit 22 degrees of bank we started to get into a buffet. That angle was the limit of our turn based on our weight and altitude. That buffet told us it was maximum performance. Kinda weak, but physics is what it is.

Abruptly, the E-Dub screamed out "Bloqueo de misiles! Romper la izquerdia!"

You see, when the E-Dub got excited he reverted to his mother tongue. English was his second language.

The Radar Nav and I looked at each other in confusion while the Pilot calmly remarked, "What was that?"

"Rastreador de meta, la una en punto!"

"Dude, we can't understand you."

The E-Dub stayed silent because he couldn't hear us. You see, E-Dubs have many sounds pumped into their headsets to listen to enemy radars. We call it the "beeps and squeaks" because that's what it sounds like to us lesser mortals. Our E-Dub's problem was his beeps and squeaks were really loud, and drowned out everything else, so he could not hear us calling him.

"Nav," the Pilot called to me. "Fix him, please."

In response, I picked up my basic checklist, a chunk of plastic and paper that weighed about two pounds, and chucked it up and to the back of my ejection seat. I had performed this complex maneuver hundreds of times and executed it to perfection because the checklist struck the E-Dub directly in the back of his helmet.

He turned towards me, startled. "English!" I screamed.

"Target tracker, one o'clock, break left!"

"Aw, shit!" the co-pilot was not pleased.

If you recall in my explanation of physics earlier, we were already in a maximum performance turn to the left at a meager 22 degrees of bank. This fact was lost on the co-pilot who immediately took the plane to around 75 degrees of bank. We went right through the buffet and into an accelerated stall. The nose sliced earthward and everyone who knew what was going on pissed themselves. We were no longer flying - we were now falling.

Obscenities were spoken at high decibel levels and higher pitches. Equipment that previously rested peacefully was thrown about. The altimeter went from 39,000 to 35,000 and then 30,000 much too rapidly. My hands went to the ejection trigger ring between my legs; assured that with one pull I could catapult myself out of the belly of the stalling aircraft. I saw the altimeter pass through 28,000 feet and gripping my ejection trigger ring I made a call to the pilot, "Do you have it? If not I'm leaving!"

As calmly as if he were ordering lunch, the Pilot's voice said to us, "I got it."

The wings leveled. The speed came into the normal range. The nose began to rise towards the horizon as I saw the altimeter bottom out at just under 20,000 feet before the needles started turning in the positive direction. The power came back in and we started climbing. I glanced at the radar scope and saw we were right over central Bagdad. "This can't be good."

The E-Dub was going, quite literally, insane. We were within range of, well, everything and everyone who was shooting. I looked over to the Radar Navigator and he was sitting bolt upright in his seat, with a death grip on his ejection trigger ring like no other. I put a hand out and touched his shoulder, startling him.

"It's cool, man. We got this. We're good."

He nodded, let go of his trigger ring and started clearing off his table. What I didn't know was in all the confusion, his intercom cord had come undone. The Radar Nav was not hearing anything either. When I got his attention, he did not hear that we were good. His repeated attempts at asking were all met with silence. He thought I had reminded him to clear his equipment before ejecting. When he pushed in his table and then grabbed his trigger ring again, I almost shit myself. "This guy is going to eject. Right now!"

"No, No, No!" I called out as I waved my hands in his face.

I reached out and grabbed his mask showing him that his intercom cord was undone and watched as he plugged it back in.

"We're good?"

"Yeah bro, we're good."

The E-Dub was still going crazy in Spanish. We had lost contact with that target tracker; we had no situational awareness on where the hell that missile was; and everyone in Bagdad decided it was time to try and shoot down the big freaking bomber. But, yeah, we were good.

We climbed for what seemed forever. I tossed out a heading to the pilot to get us out of there and then noticed that the altimeter was climbing past 40,000 feet. “Pilot, our assigned altitude was three-nine-oh.”

“I know.”

“You’re still climbing.”

“I know.”

“Are you going to stop?”

“No, I’m not.”

I heard someone call out, “Bagdad in broad daylight. What the hell, man. This ain’t Linebacker.” The pilot eventually leveled off at 45,000 feet, more or less, when he ran out of energy and physics determined he was going to climb no more. There was a short argument with the AWACS about our assigned flight level, but it was mostly ignored. We turned north and headed towards Mosul.

“Co, Radar.” The Radar Nav was calling to the co-pilot.

“Go ahead, Radar.”

“Look out your side at the weapons rack and tell me what you see.”

“Oh shit man. There’s a bomb hanging by the back lug. It’s flapping around man.”

“That can’t be good.”

“I’m going to jettison it now.” The Radar Nav pushed some buttons and flipped some switches.

“Is it gone?”

“Nope. Still flopping around.”

“Ok,” there was an uncomfortable pause. “We got a problem.”

“Have your problems later, we still have work to do.” The pilot reminded everyone. We still had a stick of unguided MK-82 bombs in the belly to drop on a ridgeline outside Mosul. Something about fixed fighting positions. Who cared, it was an excuse to carpet bomb.

We made some radio calls about our impending weapons release, basically warning aircraft in the vicinity to get the hell out of dodge. A flight of F-14 Tomcats called back.

"Are you guys going to drop a whole belly full of bombs at once?"

"That's the idea."

"Uh, can we come watch?"

"Sure." Ten minutes later two F-14s slid into position on our right wing.

"Hey, can I get under your wing, how close can I get?"

"The bombs are coming out the bay, so don't slide underneath our fuselage for any reason."

"Cool. I'm filming this."

"Go for it."

One of the F-14s stayed out on the right wing, but the flight lead drifted closer in, slightly behind and slightly low on the wing. We were lined up on our target, everything was solid. The bomb bay doors opened and one-by-one, 27 MK-82, 500-lb unguided bombs fell.

"Dude, that was so cool." The F-14 pilot remarked to us as he drifted back out towards relative safety. "If you give me your email I'll send you a link to the video." I passed over my email address and before he broke formation the flight lead remarked, "You know you got a hanger on your right wing?"

"Yeah, we know. It's stuck there."

We continued north, past Mosul into and through Turkey. We met up with a KC-10 over the Black Sea between Turkey and Bulgaria for a short top off to cover a long expected hold time at Fairford. It rained a lot in England, and the runway there was relatively short. The B-52 braking system was shit from the first day it flew and has not changed in 60 years, so we had planned on waiting to land until the rain stopped and the runway dried off. When we closed to pre-contact, the boom operator said to us, "Hey, you know you got a bomb flapping around on your right wing?"

"Yeah, we know, it's stuck there."

The Radar Nav said, "Can we try another jettison?"

The pilot remarked, “You want to jettison a 2,000 pound bomb over friendly territory while in close formation with another aircraft?”

“Yeah.”

“Ok, give it a go.”

“Wait a sec,” the boom operator called back.

“You want us to back out first?”

“No, I want to get my camera.” There was a pause. “Ok, go for it.”

The Radar Nav flipped some switches. Nothing. I tried the alternate jettison sequence. Nothing.

“You guys gonna do something or what?” The boom asked.

“Nope, it’s stuck.”

After getting gas we traversed through Europe. We passed through the Czech Republic east to west. The autopilot was flying and was supposed to make a lazy left turn to go through Germany. I had a laptop computer jacked into a handheld GPS that showed a God’s eye moving map display, saying we still had about ten minutes before we crossed over into Germany. With everything looking good, I went back to my chicken wings.

“Nav, where are we?”

“Czech Republic. About to go into Germany.”

“You sure?”

“Yeah.”

The Pilot next called to the E-Dub. “Hey E-Dub, any signals out there?”

“Nope, scope’s clear.”

“Nothing at all, no fighters or nothing?”

“To be honest, my equipment is off.”

“Well, that’s super,” the pilot laid on the sarcasm. “We got a pair of Polish MiG-29s off the right wing because you jackasses flew me into Poland.”

My greasy chicken wing fingers flew over my laptop, finding that it had frozen 50 miles back. Shit. We're in Poland. "Uh, pilot can I get a left turn to a heading of two-seven-oh?"

"Assholes. I want a new crew. I hate you guys."

The Polish MiGs escorted us to the border after several calls on the guard frequency finally got us back in contact with air traffic control. Right at the border, the MiG flight lead said, "There's a bomb flapping around on your right wing, you should fix that."

"Yeah, we know."

"Wait until you get to Germany to fix it, don't fix it over Poland."

"Understood. Thanks for the escort, sorry about the navigation error."

The hung bomb was out of the bag at this point. Admittedly we were trying to hide it or else those nations would have denied us overflight. No one really wants a hung bomb flying over their homeland. We argued with German air traffic control but eventually got our way when the Pilot just said, "Screw this, declaring an emergency."

Those were the magic words, and we got an immediate vector back to Fairford. 45 minutes later we were getting yelled at by the Group Commander over the radio as we orbited 15 miles out on final approach.

"Are you sure the bomb is stuck? It isn't going anywhere?"

"Six jettison attempts, sir. The back lug won't budge."

"Here's my concern. When you put out the chute, the sudden deceleration is going to toss that bomb off the wing. Is it armed?"

"The arming lanyard is still connected." The bomb was right next to the co-pilot so it was easy to see the lanyard still connected. When you jettison, it drops the lanyard and the bomb never arms. In this case, that didn't happen so if the bomb comes off, it arms.

"There are 2,500 protestors at the end of the runway. When that bomb comes off the wing, it's going to skip off the runway and land right in that protest."

There was 30 seconds of silence, then the pilot asked, "I don't have many options sir. You want me to dump the plane in the channel?"

"No," came an immediate response. "Can you hold the chute?"

"No can do, sir. 9,000-foot runway, wet. No way." Another 30 second pause. "Sir, we're coming up on bingo." Bingo is a fuel state that requires you to stop screwing around and put the wheels on concrete. "I was told that a divert to Mildenhall or Lakenheath was out of the question due to political reasons."

"Correct." There was another 30 second pause.

"Land the airplane. Right here, right now."

"Roger that."

The approach was flown with the co-pilot staring at a bomb. When we touched down, the pilot did not come on the brakes, he wanted to use as much runway as possible to decrease the amount of deceleration on the bomb. But when the chute came out, we were all tossed forward in our straps, as usual.

"It didn't come off! It's still there!" the co-pilot sounded surprised.

We stopped on the runway, with about 100 feet on concrete left, and were quickly surrounded by fire trucks. As we gathered under the hatch, unloading bags, I asked the crew, "Should we tell the boss about almost getting hit with two SAMs, stalling the plane over Bagdad, having F-14s in illegal formation, and violating Polish airspace?"

The pilot said decisively, "No. We will kindly omit those facts from debrief."

The protesters were about a football field away, right off the nose, as were the TV cameras. We didn't know it, but were being broadcast live over CNN and BBC. They ran a story about us taking battle damage (hence the fire trucks) and how we were heroes for saving the airplane, when in reality, the plane was fine and we were idiots.

Regardless, the Butchers of Mosul lived to fly again.

The author (left) helps apply a bomb sticker to the aircraft.

Saving Lives – Why We Train
Jim Wuensch

Another day in paradise! I loved Diego Garcia and I always eagerly volunteered to come back to the tiny atoll. It was my fourth combat deployment as a B-52 Radar Navigator. It wasn't so much the location, but the atmosphere - fly missions into bad guy land, rest, and do it again. Instead of alert, made-up additional duties, and mind numbing ground training I finally got to do what I practiced to do for two decades.

I knew this would be my last deployment. As a major passed over six times for promotion, I was coming up on mandatory retirement in nine months. It was also looking like the B-52s would soon be ending operations over Afghanistan. Things were really starting to slowdown. The daily sortie count was down and very few crews ever dropped anything. Most sorties consisted of flying low-altitude, preplanned routes in what was called "show of force" missions.

I was twice the age of my fellow crew mates. They were all quite capable at their positions though and I enjoyed flying with them. My crew's regular navigator had to leave during mid-deployment and was replaced by a brand new one right out of the school house. We only had one flight together up to this time. He probably didn't have more than 150 hours in the jet. The leadership always gave me the FNGs, but I did my best not to beat them down too severely.

Today it looked like more of the same as we had been doing before. Our crew had a show time of 1800 local to fly a standard 18-hour mission. Sortie durations consisted of a six-hour cruise to get to and from country and a six-hour "vulnerability" period in country. When you added prebrief and debrief times it was usually a 24-hour span from departing to returning to quarters. That was fine by me - I would be returning just in time to start drinking and playing poker with my honyocks.

The mission prebrief was usually quite uneventful. A cursory review of the normal ingress/egress routes, weather, and current intel. Unlike our brave predecessors in Vietnam, we faced no enemy counter-air threats. I never worried about not making it back home. Today was a little different; however, because just as we were about to step to the jet two maintenance troops entered the room and asked to speak to the RN. After identifying myself to them I sensed they were quite stressed. They told me they had been preflighting our jet and found my seat was stuck in the full-up position. They could not fix it in time for takeoff. Since I am 6'4" tall, I hated flying with my seat in anything but the full-down position. It was very difficult to see my screens with my head so far above them. I could have easily declined and chosen to move to the spare jet, but I knew that would be a big hit for the maintainers to take. Since it seemed most likely it would be another day of just boring holes through the sky, I told them I would take it "as is." They smiled in relief.

Preflight was the same as most training missions except for the weapons load out. All aircraft were loaded with 27 MK-82s in the bomb bay and 12 GBU-31 JDAMs on the wings. Because of the infrequency of actual weapon releases, the same weapons remained on the jets for extended periods. That caused a problem during power application because the cannon plug connections would corrode and caused many JDAM faults during the ground check. In past deployments, the rules of engagement (ROE) allowed taking off with some JDAMs in a no-go status. Usually once airborne the faults would clear. The rules had changed on this tour and you had to roll to the spare if any weapon faulted. That day I was lucky—all weapons good.

All operations from takeoff to getting in-country were as normal and boring as always. Most of the other crewmembers usually made a trip to the bunk for a nap or slept in their seats. In my 5,000 hours in the airplane, I never slept a wink. For some strange physiological reason, I could not sleep in any type of moving vehicle - planes, trains, or automobiles. We did check in with the Air Operations Center after tanking, and they reported no current activity in country. It looked like it was going to be another quiet night.

Once we got in country we flew a couple of the preplanned "show of force" routes before flying to the east end of the county to orbit and wait. Most of the action I had experienced in the previous four years was in that neighborhood - especially along the Pakistan border. After three hours of burning gas all was quiet until the Ops Center gave us a call and reported Troops in Contact (TIC) in a small village northwest of Kandahar. The pilot immediately yanked the jet west and firewalled the throttles.

The crew got pretty juiced, but I'd been through this exercise several times. Usually by the time you got in contact with the Joint Terminal Attack Controller (JTAC) nothing was going on or things were already over. It was nearly 300 NM to the given coordinates so we wouldn't be in contact with the JTAC for at least a half an hour.

That day things were going to be different. In every radio contact I'd experienced with JTACs in the past they were always cool, calm, and professional no matter how hairy the situation. On initial contact, our JTAC sounded like he was calling from Hell. His reconnaissance patrol stumbled upon three insurgents. At first, the insurgents fired their weapons and fled. Fifteen minutes later, however, a rapidly assembled squad of two dozen rebels attacked. One of his troops was killed. His shrill, profanity laced voice left no doubt he wanted our bombs as soon as possible. Unfortunately, he had a problem almost as serious as the Taliban - his "plugger" was broken.

A plugger is a device that uses GPS and a laser range finder to determine the coordinates of a target. Critically important for us since we had no targeting pod and could only release on coordinates. The JTAC said we would have to attack off references to his own position. Hearing this, my heart started palpitating. This would be in clear violation of all directives. The Special Instructions, or "SPINS," specifically forbade the release of any weapons when JTACs passed their own position. There had been too many incidents in the past where friendlies were killed and the guys on the ground were all very aware of the danger. I knew the repercussions in dropping could result in my courts martial, or much worse, - dead good guys. So by law, I should have said sorry and flown home. Yeah right. I threw the rule book in the honey bucket.

Official USAF Photo

The JTAC passed us his coordinates and said he wanted two 2,000 lb JDAMs at 200 meters on a true bearing of 050 off his position. That was frighteningly close. So close, in fact, the JTAC was required to say "danger close" and pass his commander's initials before we could drop. He did so without hesitation. Then it was time for me and the crew to get to work. At that time we carried notebook computers at the CP, EW, and RN stations. Among many other features, these computers could resolve coordinate conversions. Using the JTAC's position and vector, I resolved the target coordinates. I read them over the interphone and asked the CP and EW to verify. After what seemed like an hour, they finally did confirm. I then entered them into the B-52 Offensive Avionics System, but with my seat being stuck full-up made it was very difficult to see my screens. I made sure the nav verified things before the final entry. After making our turn the release point, we made our one minute out call to the JTAC. He immediately screamed "cleared hot" and before he unkeyed his mike we heard him yelling at his guys to get their heads down. When we entered the launch acceptability region I gave the "wings level" command to the pilot and pushed the launch button that's located between RN and N position. I felt the familiar "thunk, thunk" of the JDAMs coming of the pylons and waited in agony, sweating nickels.

Almost immediately after weapons impact, the JTAC came up on the radio and said in a strained tone "Okay, okay, now I want another two weapons 100 meters on a bearing 290 from the last target." The crew executed the same verification and entry routine. As we approached the release point, I looked over at my baby Nav. As he looked at me his eyes opened to the size of silver dollars and he yelled "CAN I PUSH THE BUTTON?" Talk about comedic relief! I started laughing and only made a hand gesture for him to go ahead. The weapons were released and again we waited.

After impact the JTAC quickly shouted "Good bombs, good bombs!" But with almost no hesitation, he gave us a range and bearing for another release from that impact point. He described this target as a building. My head was really starting to hurt now. I was beginning to fear we might be going full circle and dropping on his position. As a crosscheck, I used his coordinates and the coordinates of the building and determined the distance to be over 200 meters - good to go.

After we released on the building things got very quiet. After 10 minutes, I began worry we lost radio contact, or worse, so I asked the JTAC for a radio check. His only response was "standby." Twenty minutes later we were coming up to the end of our vulnerability period and bingo fuel. I was getting ready to call again when the JTAC called us. His tone of voice made him sound like a completely different person. He now sounded cool, calm, and downright happy. Our last release wiped out all the remaining enemy and life was good. He very graciously thanked us for our work and released us to return to base.

As we coasted out over the Indian Ocean on our return to the island, I started to write down as much as I could remember. I knew our debriefing would last for hours and would be attended by half the group. But it was quiet and my blood pressure was starting to fall to a level where I no longer felt my eyeballs were going to pop out of my skull. Best of all was the immense feeling of content that was beginning to overwhelm me. Absolutely nothing that I had done in my career up to that mission meant anything anymore. That day I saved lives and all the medals and promotions in the world could match that feeling of accomplishment.

Chapter Six – Bar Stories

Bar stories [bahr] [stohr-ee] – *noun* - a narration of an incident or a series of events or an example of these that is or may be narrated, as an anecdote, joke, etc. told at a counter or place where beverages, esp. liquors, are served to customers.

Odds-n-Ends

Mike Brinkman

Two of my favorite aviation quotes are by aviator and author, Ernest K. Gann:

"The emergences you train for almost never happen. It's the one you can't train for that kills you." (Or the one for which you don't train adequately.)

"Electronics are rascals, and they lay awake nights trying to find some way to screw you during the day. You could not reason with them. They had a brain and intestines, but no heart."

Another favorite is attributed to a Brit, Capt A. G. Lamplugh: *"Aviation in itself is not inherently dangerous. But to an even greater degree than the sea, it is terribly unforgiving of any carelessness, incapacity or neglect."*

Finally, the old unattributed chestnut: *"Flying is hours of boredom punctuated by moments of stark terror."* Pappy Boyington used it in his writings, as did others.

Anyone, especially those involved in military aviation, who flew for a living understands these sentiments. We all have our "war stories." Unlike many of my comrades, though, I am a Vietnam Era veteran, but I am not a combat veteran. I did, however, serve as a Cold War Warrior and have around 1,700 flying hours in the B-52F, G, D and H-models, in that order. Most of those hours were in the "Dog" model over a five-year period when I flew out of Andersen AFB, Guam. Those flights were before the B-52D received the Digital Bomb-Nav System (DBNS). All that capital was spent only to have the "Dog" model retired to the boneyard at Davis-Monthan AFB, Arizona, or to become a "plane on a stick" (static display) shortly thereafter. It is strange to know I have flown in every one of those static displays which were still in service between 1976 and 1981.

Obviously, not all stories are of the "There I was at 30,000 feet inverted when the engines flamed out!" variety. Some old crewdogs will read these vignettes and say to themselves, "Yup, been there, done that!" So, here are some of my odds-n-ends.

A Winter Day over New Mexico

In the mid-Seventies, the Andersen AFB B-52D crews did all their training over water. In order to gain terrain following low-level training, our crews would go on temporary duty (TDY) to Dyess AFB, in Abilene, Texas. This was before we had a low level route over Korea. It was winter by the time our crew, under the command of Capt Wally Herzog, was scheduled to do the Dyess thing. In fact, it snowed while we were there, which was a bit rough on us tanned, thin-blooded island boys. Even so, we accomplished all of our sorties, but there was one flight in particular I'll never forget.

It was a typical daytime training profile. As I recall, Capt Jim Bueto was the RN (though I wouldn't swear to it). We had a Dyess Instructor Navigator (IN) on board who was a crusty, yet humorous, major getting ready to retire. We were on the high level portion of the mission when the pilot declared we had a mid-body fire warning light and we needed to strap in. With all of the JP-4 jet fuel in the mid-body fuel tank, our situation could go to Hell in a hand basket in a hurry. The IN, cool as a cucumber, peered through the port in the bulkhead door and said that he saw neither flame nor smoke. It was then the major volunteered to crawl out on the alternator deck and see what was amiss. The pilot let the world know our situation and gained clearance for a descent to an altitude where we could depressurize. At that point, I realized I had taken my flight jacket off and had strapped in without putting it back on. The idea of punching out into the winter sky and landing in the snowy New Mexican desert without a winter jacket added another layer of anxiety to my situation. Of course, first I'd have to survive the ejection, descent, and parachute landing fall (PLF), but for some reason that wasn't my first concern. Then, all of our attention went to the major and his findings. The RN and I were relieved when he came back through the bulkhead door. Once on intercom he reported a bleed-air line had come loose and was blowing hot air toward the mid-body fire sensor.

Relieved, the pilot got clearance direct to Dyess and we landed without incident. The major did not have to buy any drink at the bar that night.

A Touch Too Close

Harkening back to an earlier time, after Combat Crew Training School (CCTS), I was assigned to the 28th Bomb Squadron at Robins AFB, Georgia, as a spare navigator. I had graduated from Castle AFB in the last F-model class. All CCTS classes afterwards flew the G-model as an integrated crew and reported to their assigned combat units as such. Because I arrived at Robins as a solo navigator, I was put on backburner (spare) status. Since the bomb wing was a tenant unit, I was assigned as the Wing Civil Engineering Liaison, a job I came to really enjoy.

In order to acquire some G-model training, I'd fly along on crew sorties and get a celestial nav leg here and a bomb run there. One particular sortie was remarkable, even if only for a scary moment.

It was a night flight that began as a three-ship minimum interval take off (MITO). We were number three in the line, and I had the pleasure of sitting in the IP seat for launch and air refueling. Takeoff and departure went as advertised. Next we rendezvoused with our tanker. As I recall, it was a moonless night and the A/R director lights on the belly of the tanker shown especially well. We had just moved into the pre-contact position when it happened. As the tanker's boomer swept the boom across the front of the bomber, I watched the nozzle barely graze the center cockpit window. I could hardly believe my eyes, but there, for several seconds, was an iridescent mark where the nozzle had touched the window directly in front of me. The pilots, concentrating on maintaining contact position, did not see it and I kept my mouth shut until after the refueling leg was complete. When we considered what could have happened had the boom punched through the window, we thanked our lucky stars!

The rest of the sortie went well, and the AC reported the incident during debriefing. I'm sure the word got back to the tanker squadron, and, especially, the boom operator.

Tip Tank Troubles

When I was stationed on Guam from 1976 to 1981, my favorite Deputy Commander of Operations (DO) was Col Sid Hannah. Sid Hannah was a Crewdog's Crewdog and a pilot's pilot. He took good care of his people.

One beautiful tropical day, Col Hannah went on a training mission to get in some stick time for currency. Now as I recall, the south end of the Andersen AFB runway sloped up and was above the surrounding terrain. Another factor to consider in this story is that the huge D-model tip tanks were originally designed to be jettisoned when empty when flying an Emergency War Order mission in order to reduce weight and drag. The G- and H-models had smaller, but permanently attached tip tanks.

So, during the landing Col. Hannah had the D lined up with the runway and was about to cross the overrun when one of the tip tanks made an unscheduled release and plowed into the embankment short of the runway. Fortunately, Col H. maintained control and executed a safe landing. I don't remember the incident findings, but it was probably a short in the electrical system. Gremlins?

About that time, the 43rd BW, which was short on qualified pilots, began a grand, though temporary, experiment and I became Andersen's first Supervisor of Flying (SOF) with navigator wings.

On SOF duty one day, there was another situation involving a B-52D tip tank. The BUFF crew was preparing to depart on a day mission, and I completed the drive around inspection of the BUFF and then drove to a safe position to watch the launch. The pilot was cleared to take the runway, and, as the right wing swept across in the left hand turn, the tip tank took out the runway light box on the edge of the hammerhead. The co-pilot saw what happened, and notified the crew. Thinking that the tank may have been punctured, the pilot shutdown the engines and had the crew evacuate the aircraft.

Thank God the tank did not leak because the electrical wires had detached from the smashed light unit. As you can imagine, the fire department and nearly every O-6 on base came out to "secure" the crime scene. Another close call was averted.

Ready, Aim…

The urinal on the B-52 is a tall metal cylinder with a domed top and a flip-up lid designed for male crewmembers and located at the Instructor Nav (IN) position. The bomber crew chief's responsibilities are wide ranging when it comes to maintaining his/her airplane, including emptying the urinal after each mission.

One apparently exasperated crew chief tired of cleaning up crewmember short comings around the urinal posted the following sign over the offended receptacle:

WE AIM TO PLEASE
YOU AIM TOO PLEASE

Major Opps!

All who served nuclear alert on Guam will recall the three-story building that served as the alert facility was not co-located with the alert aircraft. It was a bit of a drive to get to the sally port. On one particular tour around 1980, the crews had already endured their practice klaxon exercise for the week, so when the horn sounded at 0300 one night, it was quite a surprise!

We responded to the plane and cranked engines as the EW and I decoded the message. To say that my blood pressure elevated when the message decoded as Actual and not Exercise is accurate. But hold it! This actual wartime message decoded: "Report to aircraft - Do not start engines!" The klaxon had sounded and thus dictated for the crews to report to aircraft and start engines!

The command post radio frequency lit up with queries from each alert bird for a repeat of the message and further instructions. "What do we do next?" All we got from the command post initially was "Standby!" Apparently they were confused, too. Finally, after about 10 nerve-wrecking minutes, we got the word to shutdown engines and return to the "cocked" status. Needless to say, we didn't get much sleep the rest of the night.

When we finally got an explanation for the SNAFU, we were told a capacitor in a computer at SAC HQ malfunctioned and set off the "fast klaxon" worldwide. We wondered just much JP-4 the Air Force used up vis-a-vie the cost of one capacitor.

Bomber Explosion

Even though we Crewdogs did not always act like it, we did respect the hard work our maintenance troops performed to keep us flying. But I don't think I often considered how physically demanding and even dangerous working the flight line could be until 27 January 1983.

At that time I was the Emergency War Order (EWO) Training Officer for the 319th Bomb Wing at Grand Forks AFB, ND. I was working in the vault, which was located in the Alert Facility on that tragic day when we got word a B-52G had exploded on the flight line. I joined a large group of alert crewmembers and maintenance personnel outside to witness the horrifying events unfolding a mile away. We watched as brave ground and flight personnel taxied other bombers away from the burning hulk. Five maintenance troops died trapped in the burning fuselage. Eight others were injured. The accident investigation concluded someone on the maintenance crew kept resetting the circuit breaker for the fuel pump in an empty tank. The pump overheated and ignited the vapors in the fuel tank.

The entire wing was in shock and grief stricken. Two of the men who died were married, one with a little girl. The day after the accident I was assigned as the action officer for his family. It was a great honor to serve this family in such a difficult time. With the help of the Operations Maintenance Squadron commander and first sergeant, and the widow's father (a retired USAF major), we managed to have the young airman buried with honors in Arlington National Cemetery. We then got the family packed, cleared their quarter's inspection, and on their way to the widow's parents home in Spokane, WA. All of that was accomplished in 10 days. There is no doubt the US Air Force takes care of its own, especially in times of tragedy.

In Memorial

I guess you could say that I'm a second generation BUFF flyer. My dad, Fred Brinkman, was a field engineer for IBM. He had assignments taking us to Boeing bomber plants in Wichita, Kansas, and Seattle, Washington, as well as SAC bases in Lake Charles, Louisiana, and Moses Lake, Washington, where he performed flight tests of the bomb-nav systems on the B-47 and B-52. Later, we moved to Eglin AFB, Florida, where he flight-tested the Hound Dog and Skybolt missiles' crew-to-missile interfaces. I don't know how many flying hours Dad had, but I do know that he was proud of his contribution to the SAC mission.

Sadly, Dad died in 1974 at the age of 47. I was attending Combat Crew Training School at Castle AFB, Atwater, California, at the time. I never did get a chance to compare notes with him about our experiences with the BUFF BNS.

One Long Day

Russell Greer

While assigned to the 62nd BMS, 2nd BMW, Barksdale AFB, Louisiana, in 1986, our crew was in the middle of another fun-filled alert tour when we were summoned to the vault. It seems as though ole Muammar Gaddafi was acting up again and was linked to some recent bombings in Berlin and President Reagan had let him know we meant business in April with airstrikes. There was a chance our unit would be tasked with a "Show of Force" mission within the next couple of days and the wing staff wanted to ensure we would all be ready if we were needed. It turned out that we were not called upon while we were still on alert so we went into our normal Combat Crew Rest and Recuperation (CCRR) or "C-Square" as we normally called it.

We kicked off the following week with a full day of mission planning for a sortie to the Strategic Training Range Complex (STRC) and entered crew rest. Shortly after entering crew rest we were called back into the squadron building along with two other senior crews and an extra pilot and NAV for each crew. The briefing began with the Wing King and his court letting us know this was a planned show of force with a three ship B-52 cell to Gaddafi's "Line of Death" and return. The flight would also test an experimental Voice AFSATCOM system at the gunner's station. After the dog and pony show ended we went back into crew rest with a briefing show time scheduled for the following afternoon.

The sortie had a planned duration of 32 hours including three hours of low level activity along Gaddafi's "Line-Of-death." If memory serves me correctly we were scheduled for three air refuelings. The plan for the VAFSATCOM test called for hourly position and SITREP reports which meant little or no rest for the gunner. Our crew planned to eat chili dogs from the hot cups along with the normal in-flight kitchen meals.

We arrived at the aircraft the normal hour and a half prior to take-off and found all three aircraft were in the green and preflight and departure went smoothly. The first few hours passed just as any other sortie with the VAFSATCOM performing as expected. Shortly after completing the first scheduled A/R (somewhere off the coast of Maine) the pilots noted low oil pressure on one of the engines but we continued the mission. It was probably at about the 10 to 12 hour mark when we all first realized it was going to be a long day. The BUFF cockpit was beginning to get cold soaked and we were just hitting the one-third mark in the flight.

Sometime after the sun began to set the oil pressure reading that had been low for a few hours took another drop and the pilots decided to shut down the affected engine and continue on with just the other seven. The NAV was taking his Cell shots (Celestial Navigation for the crew newbies) and we were preparing to conduct the Bonus Deal Bomb Run scheduled for the sortie. For those not familiar with the Bonus Deal procedure it was used in case one of the aircraft in a cell lost their Bomb-Nav radar prior to reaching the target. The gunner in the preceding aircraft would furnish range and degrees left or right of center line, and that information was used to know when the following aircraft should release its bombs. The Bonus Deal would prove its worth once again during a sortie with the 806th BMW (P) years later during Desert Storm - but that is a story for another time.

If you have ever had the pleasure of a long flight from west to east over the Atlantic Ocean you know that the night is short. It seemed as though the sun had just set when we saw dawn approaching from the east. The seemingly short night along with hours in an uncomfortable ejection seat seemed to really set the fatigue in place. As we entered the Mediterranean through the Strait of Gibraltar we all seemed to perk up a bit. Not only were we getting close to our descent to low level but it was a beautiful day in the Med (at least for the crew up front that can see outside the cockpit.)

The mission plan called for a descend to low level and to cruise up and down the Libyan coast to see if we prompted any sort of reaction. Our planned duration for the low level portion was approximately two hours. We had "hot" guns just in case we needed them. This is a situation to which everyone knows the ending. Gaddafi spoke loudly but was all hot air. After two hours of the EW calling out nothing but search radar signals, we climbed out and headed for home.

The touchiest part of the flight occurred on the way home. As I noted, the weather in the Med was beautiful, but it did not continue to be as we made our way back into the Atlantic and our third and final A/R. A front was producing thunderstorms over a wide area and the planned A/R track was going to be smack in the middle of it. With some juggling of altitudes and routes the pilots and tanker crews found a somewhat better track for us and we were able to get the gas in the weather with only a few disconnects. It was a bit touch-and-go for number three in the cell and took him a bit longer for but he did eventually manage to "fill'er up." The air refueling skills demonstrated by the majority of the SAC pilots I flew with really stood out later in my career when I went to flight engineer school in 1992 at Altus. My first A/R sortie as a student F/E on C-141's with a struggling pilot upgrading to instructor made my appreciation for my BUFF A/Cs grow even more. True, I flew with a number of 141 A/Cs that had mastered A/R, but many struggled with it.

Once our final refueling was over it was just a matter of logging hours to get back to Barksdale (KBAD). The fatigue was really setting in for everyone by then and at about the 26+ hour point I crawled out of the seat and laid down behind the IP seat to nap about 30 minutes. It seemed as though I had just dozed off when the EW gave me my wake up call for another SITREP on the VAFSATCOM.

Well we finally made it back to KBAD and after taxing in and shutting down we were met by everyone on base it seemed. We were all exhausted but as any crewdog knows the sortie is not completed until the maintenance debrief is over. As tired as I was, I was not in the mood for a ton of chit-chat at the plane and as I stepped down out of the hatch someone asked me how I thought the SATCOM had performed. It was a simple question, but the wrong time. I stated that I would be happy to give all of the details at the debriefing and as I turned around I saw the person asking the question happened to be the 2nd BMW commander. He gave a small smile and said, "Long day wasn't it?" I blurted out a quick apology and gave him a report and he respond with a slap on the back and a "Great job and don't worry about it."

Obviously there have been longer B-52 sorties and I grimace when I think back to that three-ship around the world flight that lasted 45 hours. Not only did they log an additional 13 hours but one of the gunners stayed in the tail the entire time. I had the privilege of serving with that gunner's son, Joe Preiss, during my career. Joe was the true "Son of a Gunner" as well as going on to become a flight engineer stationed at McGuire AFB after we were pulled from the BUFFs.

As I write this story, 25 years after stepping out of the hatch from my last BUFF flight, it seems as though the time has passed in a flash. I had the honor of serving with some awesome crewdogs, ranging from my first instructor at CCTS, Anthony Freeborn, to my flight-line instructor, Mike Riggs, to all of the great gunners, EWs, NAVs, RNs, co-pilots and ACs in the 62nd and 596th BMS at Barksdale to those at the 436th STS at Carswell.

My finial BUFF flight took place with a 20th BMS crew at Carswell in October of 1991.

My time as a SAC Crewdog were the best years of my professional career. Serving in the greatest combat command of all time, SAC, with the greatest Crewdogs ever, on the greatest combat aircraft of all time is something that will stay with me until the final Klaxon sounds.

For my fellow Bulldogs I say, "Nuthing could be funner than being a gunner!"

Sleeping Around on Guam
A Crewdog's Confession

Tommy Towery

It was a cold February night and I was enjoying a good night's sleep in my newly acquired king size waterbed when the midnight call interrupted my nocturnal bliss. Earlier in the evening my wife and I had been playing cards with another couple we were friends with when he also got a call. He worked in the 7th Bomb Wing Combat Intelligence office and was told to report to work immediately. Often he would be called in because something set off the security alarm in the vault where his office was located, so I really didn't think much about it. When my own midnight call came a few hours later that was not the case. I was told to get packed for six months and report in uniform ready to deploy at 0800 the next morning. Little did I know how much my sleeping habits would change in the months ahead of me.

My phone call started my countdown for Operation Bullet Shot. My orders were effective 9 February 1972, sending me on temporary duty (TDY) from Carswell AFB, Texas, to Andersen AFB, Guam for "Contingency Operations in Direct Support of SEA." I showed up on base the next morning, filled out a power of attorney, got my shots, made out a will, and collected $8.00 a day advanced temporary duty pay for 180 days – $1,440 in cash. Though I had first been assigned to Carswell as a B-52 Electronic Warfare Officer (EW for short) at the time of the call I was medically grounded for kidney stones from flight duty and serving as the Wing's Assistant Penetration Aids Officer. I was a staff weenie.

I wore my flight suit when I boarded the contract DC-8 aircraft and headed for "The Rock." It was the first time I had ever flown on a commercial aircraft. I am sure I did not get a full night's sleep but only cat napped on and off during the trip. Only one incident about the journey stands out in my memory. One of the cute flight attendants (we called them stewardesses back then) was paying special attention to me because of my captain's bars and the wings and colorful patches on my flight suit. Most of the other 100 plus airmen on the airplane were enlisted maintenance personnel and were wearing their drab fatigues. Several times early in the flight this stewardess would check in on me and we chatted about different things. She started talking about the patches I was wearing and found the yellow and black eagle-in-flight patch of the 7th Bomb Wing especially intriguing. She held my shoulder with the patch and sweetly asked what the Latin "Mors Ab Alto" wording meant. I replied it translated to "Death From Above" in English. She let go of the patch like it was a snake, pulled away while frowning viciously and blurted out, "That's terrible!" I never saw her for the rest of the trip.

I cannot remember how long the flight took, but I will always remember how tired I was when I arrived at Andersen. After a not-so-short trip through an in-processing line, I and my fellow travelers were taken by bus to our sleeping accommodations for the night. The base gymnasium had been transformed into one very large dormitory room with wall to wall cots. It was first-come-first-served finding a bed for the night, and the only good thing about it was we were all so tired it did not matter how loud people were snoring or what else was going on around us – we slept.

Approximately 8,000 Air Force personnel were sent TDY to Andersen under the guise of Bullet Shot. Someone came up with the saying, "Bullet Shot – The herd shot round the world." The base was not prepared for such an onslaught of people needing living quarters. After two nights in the gym, I was finally assigned a room in one of the three-story concrete barracks just down the hill from the Officers' Club. I was assigned duty as an 8th AF Arc Light Mission Planner where I worked real shift work for the first time in my life. My schedule was three days day-shift from 0800 to 1600 followed by 24 hours off. I then went on a swing-shift from 1600 to 2400 and 24 hours off. The night-shift was 0000 to 0800 and then three days off. Sleep time was anything but normal. I was bunked up with a fellow officer from a different base who I had never met – I can't even remember his name. He was on a different shift schedule from me, which was not all bad.

With all the new crews and aircraft assigned, the bombing schedule meant Andersen operations became 24/7 with most people working shift work to keep the missions flying. My roommate and I passed like ships in the night, sometimes not seeing each other for days. He was a very hot natured guy, in contrast to me being cold natured. Even though Guam is a tropic island and military bases are notorious for faulty HVAC equipment, the air conditioner in our room was highly efficient. This led to some interesting sleep situations. I would come in from work and turn up the air conditioner and go to sleep in my bunk with just a sheet over me. A few hours after I was asleep my roommate came in from the heat, turned down the air conditioner as low as it would go and hit the rack. He was one of those who loved to sleep in a cold room under several layers of blankets. I would eventually wake up freezing and with chill bumps and get up and turn the air conditioner back to a comfortable level for me – which left him wrapped like a baked potato in a hot oven, sweating profusely. On days or nights when he went to bed the routine reversed. We never seemed to find a compromise on the temperature and I don't even remember discussing it with him, so this ritual continued as long as we were billeted together. That was not long.

The second phase of Bullet Shot came in May, when the D-model B-52 force was augmented with the G-model troops. By July 1972 a force of almost 50 B-52Ds, 100 B-52Gs, and over 12,000 personnel were bedded down at Anderson Air Force Base. I quickly found out my room was needed for the G-model crewdogs and was ordered to vacate.

I soon found myself living in a trailer on base, two officers per side with a shared bathroom in the middle. I felt safe there, because they were all secured by cables to keep them from blowing away during typhoons. Their outside appearance reminded me of what FEMA trailers look like. It wasn't so bad, but the continued influx of crews soon evicted me from my "Trailer for sale or rent."

This was the time when it paid to be a staff weenie. Since the aircrews needed to be on base and near their aircraft, they needed all the on base quarters available. Some metal buildings previously used only as storage facilities became quarters for the enlisted troops, and hence, “Tin City” was born. As more and more personnel arrived and more facilities were needed to house them, canvas tents appeared overnight like mushrooms in a field. This became “Tent City.”

The staff however had set schedules and could work out getting to and from the base without too many problems. The solution was to contract with some of the local hotels to house the overflow of personnel. The area with the most rooms available was a major tourist spot for Japanese tourists – the beautiful beachfront area of Tumon Bay. The result of this plan put me in a small bungalow in a hotel complex called The Guam Continental. The hotel was also used by many of the commercial flight crews transiting Guam and was very luxurious. Once again I shared a room with another staff officer, but like before we rarely saw each other. The bungalow was in a palm tree and tropical plant garden setting with small walkways leading between the several dozen units. Less than 25 yards away was the white sandy beach and in the evenings I could hear the ocean waves break on the shore as it lulled me to sleep. It was truly paradise, except for the loneliness and the grueling shift-work schedule I endured. A base blue shuttle bus was set up to ferry personnel back and forth from the hotels to the base and sleeping on the 30 minute trips to and from work became my normal ritual.

As my TDY ended and I headed back to Ft. Worth, I looked forward to returning to my own wife and my wonderful waterbed. And I did – but only for a short while. In September I got another set of orders and once again I was off to Guam. “Hafa adai!” again. My accommodation for this soiree was once again in contract quarters. I was assigned a room at the Fujita Tumon Beach Hotel, back at Tumon Bay. The Fujita was not the exclusive Continental Hotel I had vacated earlier, but was more like an extended living motel seen today. Each unit had two bedrooms, one bath, a small living room and a full kitchen. Two staff officers were assigned to each room, so each unit housed four officers. Again shift work meant we came and went at odd times but at least this time I got to know my fellow residents. I was roomed with Lt Bill Perry, a Civil Engineer type. Our room had a window in the back, which was often left open at night to take advantage of the tropical breeze. This proved to be a problem, because right behind our building was a Korean restaurant and the smell of kimchi being cooked early in the morning was nauseating.

The other bedroom was occupied by Capt Charles “Chuck” Massey, a maintenance officer, and a major named Bill whose last name I shall not reveal. Bill had a real drinking problem but was a fun roommate anyway. He would go to the commissary and buy some fresh fish to cook in our unit. He normally started drinking as soon as he got home and would often pass out before he cooked the fish. One of us would come in several hours later and see the raw fish still sitting on the table after spending hours out in the warm tropic temperatures. We would throw the fish in the garbage. The next morning Bill would get up and find the fish in the trash can, take it out, and cook and eat it.

On base temporary quarters back then were not as well furnished as the ones today which have coffee makers, small refrigerators, and microwaves. Having a real kitchen was a change and though I did not cook many meals, having a refrigerator allowed me to keep milk cold for cereal and sandwich meats. In my previous dwellings I normally ate most meals at the Officers' Club or at the food truck outside the 43rd Bomb Wing's Ops building. The crew name for the place was "Gilligan's Island." When I wanted something a little better, there was a place outside the back gate called Kenny's Steak House and when some of my friends had access to a "Guam Bomb" we would travel to the Navy Officers' Club. Of course, an order of French Fries from the bowling alley was always a quick fallback meal, and sometimes my meal was a box of popcorn at the "wash out" movie theatre. It was cheaper to have a kitchen.

Now comes an interesting fact about the contract hotels. Not knowing how long the TDY personnel would be staying, the base housing office only contracted the hotels for a month at a time. And when I say month, I mean four weeks. Four weeks is only 28 days, which worked out fine in February, but for the rest of the year it meant two or three days were left uncovered by a contract at the end of each month. The hotels took advantage of these non-contracted days and rented them out at high rates to the Japanese tourists, which meant the TDY personnel occupying the rooms the rest of the time had to pack all their things and find somewhere else to sleep – on base.

That was when the staff finally felt the pain of the enlisted troops. We were moved into the tents! Yes, there is nothing like sleeping in a hot canvas tent in the middle of the day on a hot tropical island when a three-ship cell of B-52 aircraft was taking off every 30 minutes less than a quarter of a mile away. When the BUFFs were not taking off, they were landing.

Tent City was my home for the last few days of September, October, and November, but the end of December 1972 was a different story. History recalls that Linebacker II was in full swing then. Very few people got any good sleep during that period. In fact, I was so busy, and the ability to sleep in the tents was so difficult, I slept at my desk or in an office corner during the entire campaign. Most of the time I was busy planning the bombing missions and with the changes coming the way they were, it felt like a 24-hour work day.

With the success of Linebacker II, many troops were rotated back to the states and I was one of them. I would not return to Guam for almost two years. By then I was back on a crew and deployed under orders for Operation SEA Surge. It is difficult to find any lasting information about that operation, but basically my B-52D crew and I were TDY to Guam and U-Tapao for three months, alternating between the two bases. When we were not pulling alert in the barracks converted to an alert facility, we were housed in the same building I lived on my first trip to Guam, two people per room.

While on alert all five crew officers were in one room, and the enlisted gunners from several crews shared a room. We slept in bunk beds, ever ready to react to the Klaxon's call. There was a common bathroom, but the open-pipe plumbing between floors allowed the noise to filter from one floor to the next. On one short trip to U-Tapao we were allowed to store our cold-weather alert flight gear in an unused room while we were gone. When we returned all the clothes were mildewed from the humid weather and the smell never went away. My gear still smelled like mold when I turned it in to go to my next assignment.

My last visit and last sleeping quarters on Guam came after I had been assigned to the RC-135 Rivet Joint program. My crew was TDY to Okinawa when we were typhoon evacuated to Guam. Again, there was no room at the inn and we were bused to an area called Anderson South, which was a World War II housing area. I remembered it from my earlier days, since it was a stop on the base shuttle bus route. Some may remember it as the Marbo Annex Housing. The facilities had not been used for several years and we were let into a dimly lit concrete barracks building with rooms full of dusty mattresses stacked in a pile. We were told to find a mattress and find a room, get some sleep and a bus would be sent for us the following day. Several of the crew members elected to go back and sleep on the plane rather than risk the threat of jungle rats. I slept on the floor on the dusty, mildew smelling, mattress.

So, that is the gauntlet of sleeping quarters I utilized on the beautiful tropical island of Guam. The Marbo housing had to be the worst (even worse than the tents or my desk) and the tropical garden bungalow of the Continental Hotel sitting on Tumon Beach the best. I don't guess I can complain since I usually had a bed and was not forced to sleep in the jungle or a sandpit. I just have to remember the old saying, "Guam is good, by order of the base commander."

Typical SAC Alert Facility

Turner AFB Reprise

Derek Detjen

After writing the first memoirs of our years at the 824th Bomb Squadron, additional fondly remembered incidents continued to resurface. Here is an additional bunch of stories I think are hilarious memories of my favorite bunch of crewdogs, during the height of the Cold War.

What Airport?

One of the worst things that could happen to any peacetime B-52 aircrew was to return from a training mission with a dreaded "bad bomb" score. The immediate future career of both the Aircraft Commander and R/N might be in jeopardy, not to mention the future promotion of the Wing Commander. At least weekly, and sometimes daily, depending upon the circumstances, a "bad bomb panel" would be held in the Wing HQ building. The Wing CO or usually the Wing Deputy Commander for Operations (DCO), also known as the "hatchet man" for the Wing CO, would conduct the interrogation of the crew in question. Also present would be a number of the wing staff, and often the bomb squadron commander.

Never was the old SAC adage of "To err is human, to forgive is not SAC policy" truer than in those hearings to find out who was at fault: the crew, maintenance problems, or other causes. One of our more light-hearted R/Ns, the inimitable Capt Dick Cole had gotten a huge bad score, and it was soon determined the crew had taken off on their mission on June 30th, but did not arrive at Statesboro RBS for their practice runs until July 1st. Unfortunately, the new bomb run routing had been changed, effective July 1st. The crew had not been given the new maps, or the new classified codes that were also superseded on July 1st. It was a blunder that ultimately targeted both the Wing Bomb Nav shop and the Tactical Communication folks as being responsible for the 99,000ft+ score. Ergo, the aircrew was not only flying with the wrong low level route map, but was unaware of the correct gross error of their bomb score!

Just before this fact became known, the erstwhile DCO, the lovable Col Willie Sonntag, already crimson with anger had Capt Cole in a brace and said "Cole, you idiot, anyone knows that even if your radar is out, you can just pass the end of the main runway at the Statesboro airport, count to ten and cut the tone, and you'll have a reliable release!" In a never-to-be-forgotten faux pas, Capt Cole replied "Airport, what airport? I didn't see any airport!" The classic response would live on in infamy in the folklore of the 824th Bomb Squadron!

Flying with the Luftwaffe

Capt Walt Costello was, without a doubt, Turner AFB's answer to Bob Hope. Walt's dry sense of humor and his constant quips made any evening movie at the alert facility worthy of attendance. Once it was determined he would be an attendee at the night's flick, often an old "grade-B" Hollywood production, the briefing room would be packed, mainly to hear Walt's latest remarks about said movie. The movies were always worth the price of admission!

Capt Costello's wife, Louise, was a super gal who grew up in Germany. Walt had never met Louise's parents since their recent wedding, so a lengthy trip to Germany was planned and Louise was sent ahead on a commercial airline. Walt went "space-A" leaving from an east coast base flying to England. From there it was on to France and upon landing, he learned MATS had cancelled its afternoon flight into Cologne due to predicted high crosswinds on the main runway at Orly. So there he was, already up for a full day and wearing a grimy flight suit and no way to get to his final destination. Finally, out of the blue, appeared his solution in the form of a German Air Force Lockheed C-119, also known as "The Flying Boxcar."

The Luftwaffe crew soon appeared in the military Base Ops building, complete with a bunch of German aviation pilot training students! Walt immediately asked the Aircraft Commander if he was flying back to Germany that afternoon. "Yes, ve are going to Cologne" the pilot answered. "Fantastic," Walt replied, "Could I bum a ride with you?" The response was "Vell, ve like to let da Amerikan capitan fly vis us, but is against the German Air Force regulations!" Walt persisted with his request, and soon a series of phone calls back to Germany finally resulted in Capt Costello getting a personal phone call from the President of the Republic of Bonn, personally authorizing him to hop a ride on the old C-119!

Subsequently arriving at the plane for preflight and takeoff, Walt watched as the entire bunch of aviation cadets stood in a brace with their parachutes on, getting a thorough inspection from the crew prior to takeoff. There was no laughing, no talking, and it was obvious that this was indeed a military operation! Finally, the German pilot gave Walt a chute, a brief checkout on their bailout procedures and asked, "Vell do you haf any questions?" Walt looked at the pilot, at the row of rigid cadets, thought for a moment, and then replied "Yes, I have one question. Why do so many of you guys look like Kurt Jergens?" Discipline was shot for the entire flight, as many of the crew and student pilots, doubled over in uncontrollable laughter! Walt's later recounting of his trip was even better.

Leopoldville

Thursday morning was "changeover day," and a bunch of the ongoing alert crew force was riding in a blue Air Force bus from the 824th Building to the alert facility. Even though it was during the height of the Cold War and nuclear safety concerns were quite stringent, people were not as paranoid as they would later become. As we circled the south end of the main runway at Turner, the guys on the bus spied an MMS tow truck dragging a huge nuclear weapon on a trailer behind him. The truck was also en-route to the alert area, obviously making a weapons changeover on one of the aircraft in the "Christmas tree." Several MMS enlisted troops were slouching nonchalantly on the trailer and a few of them were perched astride the big nuke, kind of like Slim Pickens did in the famous "Dr. Strangelove" movie of that time period.

Capt Costello eyed their progress as they rode along parallel with our crew bus, inside the perimeter fence. He remarked, "It looks kind of like a bunch of U.N. Troops entering Leopoldville, doesn't it?" The bus erupted with laughter. Walt was a great crewdog to pull alert with, believe me!

First and Last Day on Alert

It was a warm summer evening, and a new Air Police security guard was on duty on alert Sortie 01, which was parked in one of the first two parking spaces on the alert Christmas tree. Just after dinner, a devious plan was put into action involving a beautiful, small remote control model of an Air Force staff car, complete with rotating beacon, multi-channel radio control steering, brakes and a two-way walkie-talkie mounted inside drove slowly out towards Sortie 01. The unsuspecting guardian was standing in front of the "No-Lone Zone" red perimeter line. A large number of crew personnel watched from just around the corner of the alert building as the station wagon stopped in front of the airman's feet, and a pleasant voice from inside the model said "Good evening, airman, nice weather, isn't it?" The guard was wide-eyed as he quickly called his supervisor and reported the incident.

The Air Police SMSgt quickly calmed his young charge down and assured him all was well, at least for the moment. Darkness descended on the base, and the alert pad Christmas tree was silent at least for the moment. Suddenly, a muffled explosion came from inside the B-52 on Sortie 01, followed by a loud hissing noise. The guard, now extremely agitated, and fearing some type of enemy action, called a "Seven-High," and was temporarily removed from his post after help arrived. Two Sortie 01 officer crew members arrived at the aircraft, along with the crew chief and several maintenance types. It was soon determined the co-pilot's oxygen hose had broken loose from its mooring and was whipping back and forth, driven by the 300 psi of the system. Order was restored, the system was repaired, and the by-now thoroughly agitated security guard was finally left by himself to guard his aircraft as the night grew longer.

Shortly thereafter, a test ICBM was launched from Patrick AFB, Florida, into the evening sky, its exhaust leaving a gorgeous, concentric trail of red, yellow, green and blue rings in its wake as it ascended into the stratosphere. The wide-eyed guard, now babbling almost incoherently, called a second "Seven-High" and was finally removed from his post and order was restored. He was never seen again on patrol duty at the alert area and his fate is still unknown. Whenever I think of that poor young fellow, I always wonder what would've happened if he'd been on duty at night at Minot or another Northern base in a below-freezing winter environment.

Finally, I'd emphasize that any prank ever pulled while on alert duty, similar to the radio-controlled staff car was first fully coordinated with the Air Police powers that be and approved by them before the fact. The alert pad, the two-officer and two-man policies and the perimeter around each aircraft were always scrupulously enforced.

The Hobby Room

There was an unused, empty room across the main hall from the CQ office upstairs in the alert facility that soon became a designated area for all the would-be-hobbyists to use for building their latest boat, train, or aircraft model. The smells emanating from this room were quite intoxicating, with a mixture of Floquil and other paints, lacquers, and various paint thinners. Usually, about 30 minutes or so of those aromas was enough to send one into a crew lounge or the dining hall for recovery!

It was interesting to note that among the crew force, there were at least four or five model railroad fans for every radio-controlled aircraft builder or boat enthusiast. The room soon outgrew its capacity for us about the time that the alert dentist room in the downstairs sleeping quarters was vacated. This was done due to a couple of incidents when the Klaxon went off while a crewdog was in the dental chair with a drill in his mouth! Due to the number of model RR builders, this abandoned room soon became the location of an HO scale railroad empire, complete with a multi-block control system capable of running two trains at the same time without encountering any mishaps.

Capt Tom Tobin, mentioned in previous volumes of WWCD, was a known non-fan of navigator types. He happened to walk by late one afternoon, and after watching from the doorway for awhile, asked "Hey, can I come in and learn to run the trains?" He sat down at the main panel, and after some basic instructions, was soon effortlessly running a freight train around the mainline. After about an hour or so of this and with dinnertime approaching, he exited the room, paused and opined, "Yep, my kids would really like this." His remark was interpreted as a mild putdown of all the navigator types in the room!

The Traffic Pattern

Two of the multi-channel, radio-control guys who were really interested in that developing field of model flying were Maj Hal

Owens, the AC on one crew and Maj Durrell Shafer the R/N on my crew E-24. Durrell love to engage in “dogfights,” in which two competing planes with long streamers attached would try to hit the competing plane's streamer and thus “win” the fight. One afternoon, a bunch of us were sitting out on the baseball bleachers watching when the predictable midair occurred. Some $200 worth of Durrell's blue and yellow PT-17 wafted downwards in a cloud of balsa wood, electronic movements, the engine, etc., to the cheers of the gallery! After dinner that evening I passed by his bedroom and viewed a personal funeral pyre of the PT-17, arranged in a neat pyramid on the R/N's bed! Combat losses were to be expected, I guess.

Maj Hal had a huge, 15-foot wingspan single-wing aircraft that usually performed flawlessly while on alert. It was a heavy aircraft, and needed a lengthy takeoff roll before becoming airborne. On many occasions, Maj Hal launched it down the throat of the alert pad parking area and then flew it around the north end of the main runway. On one memorable occasion, with both the bomb and tanker squadrons doing touch-and-goes on the runway, the alert facility got an angry call from the Tower, requesting “The pilot of the model aircraft to clear the main runway area immediately!” Some folks just have no sense of humor.

Andersen AFB Alert Facility

Stupid Alert Tricks

Michael L. "Mike" Brinkman

Most any BUFF or KC-135 crewdog who ever sat nuclear alert every third week will admit that as they passed through security into the Alert Area for crew changeover a little switch clicked in their heads and they took on a different mindset. I call it "Alert Mentality." That is, do whatever it takes to get through the next seven days and back out to the world on Combat Crew Rest & Relaxation (C^2R^2 or just C^2).

Emergencies at home seemed to await your departure if you had a family. That's when your kid ended up in the emergency room, or the washing machine hose busted and flooded your quarters, or your spouse went into meltdown because the car died, and there you were, stuck on the Alert Pad.

Since we were a captive audience for seven days, the bomb wing staff was more than willing to help us pass the time with no-notice tests, simulator practice, various and sundry classes, Emergency War Order (EWO) sortie study, preparation for the upcoming Operational Readiness Inspection (ORI), Buy None, etc., etc., etc. Some crewmembers worked on master's degrees, leadership schools, hobbies or played games, or watched TV or movies to while away the week.

With all the aforementioned activities, one might assume there was little time left for mischief. But, oh no! As the days ground on, Alert Mentality reared its ugly head even higher. That was especially true if the weekly alert exercise had not occurred yet and or the ORI deadline was near. It seemed that as the end of the alert tour loomed, the number of pranks and tricks crewmembers played on each other increased. Usually these episodes were harmless enough, but sometimes they were downright dangerous!

Now, I'm pretty sure that in reading this, every crewdog has by now conjured up in their minds their favorite (or least favorite) alert tour pranks. It depends on who the target was: you or the other guy. Following is one prank that I will always remember.

While stationed with the 60th Bomb Squadron on Guam from 1976 until 1981, I crewed with a number of different aircraft commanders. Some were great and some were lacking. The AC, who was the victim of one particular incident, was well liked by everyone on the crew (honest!). Because he was rotating back stateside, it was our last alert tour with him. That being the case, and it being late in the tour, the crew came up with a plan to give the AC a proper alert sendoff.

Now, anyone who sat alert at Andersen will remember the Alert Facility was not a typical SAC mole hole co-located with the aircraft. Instead, it was a three-story concrete barracks with outside covered walkways running the length of the building. Stairways ran from the walkway to the ground. Also it was located about a half mile from the alert birds. The building was surrounded by a chain link fence topped with barbed wire. I don't know if that was to keep outsiders in or to keep crewdogs from leaving.

The AC was bunking with the co-pilot in a second floor room. On the night before crew changeover, the co-pilot checked on the pilot, finding him on his bed in his skivvies reading a magazine. Before his arrival, we had stripped the sheet off one of the vacant bunks and soaked it in the shower. The co-pilot left the room to join the rest of us. In the meantime, the gunner moved the alert truck to just outside of the facility gate. At 2230, the rest of the crew quietly gathered outside the AC's door. On the signal, we rushed into the room and pounced on the hapless pilot. We managed to wrap him up in the wet sheet cocoon (for a skinny guy, he put up quite a struggle!). Three of us got a secure hold on our mummified captive and marched him down the stairs head first, out the gate, and into the bed of the pickup with two guards. Thank God we didn't drop him! Before leaving the pilot's room, we snatched his flight suit and boots just in case the horn went off. Muffled cursing could be heard as we drove carefully to the Officers' Club.

The streets were pretty much deserted as we traversed the base to the darkened O' Club. At that point, we hustled the white, wiggling body bag out of the back of the truck and to the edge of the club swimming pool. After rapidly unwrapping the disoriented AC, someone gave him a little shove and into the pool he went. It was no time to be hanging around, so we all rushed back to the truck and drove out of sight before stopping and shutting down. A quiet five minutes passed before we went back to retrieve our sopping wet leader. Someone tossed him a towel, and we made him sit in the bed of the truck all the way back to the facility.

By the time we got back there, the pilot (who, by the way, was a practical joker in his own right) realized the honor we had bestowed him in true crewdog fashion with such an end of tour prank. No harm and no foul. A private, though sober, party ensued in the pilot's room until the wee hours.

That night sure made for a great memory.

Crewdog Stories

Ted Lesher

Fly and/or Fight

Although the Strategic Air Command (SAC) stood at the apex of the entire U.S. military establishment, it didn't take me long to find out that within SAC the people who actually flew were on the bottom, hence the appellation "Crewdogs." I knew a British exchange officer who couldn't understand how we had evolved a system that valued desk jobs above flying, trained every officer to one day be chief of staff. At the same time this was being done every little decision was moved to high-ranking officers carrying two-way radio "bricks," and an up-or-out policy discarded perfectly capable people who for one reason or another had failed to be promoted. I didn't understand it either, but was fortunate to make it to retirement as an aviator with very little staff experience. In the meantime, my friendly fly-off rival had his own stroke of good luck: he got kidney stones, which took him off flying status, got him a staff job, put him through advanced education, which led to a good career in the Air Force and beyond.

Heightened DEFCON

In late November 1963 three of us were going through B-52 combat crew training at Castle AFB, California. On one particular Friday the only thing we had on the schedule was an afternoon class on command and control procedures, so we asked the instructor, Maj. Don Humann, if we could take it in the morning and start the weekend early. He was a command post controller and glad to oblige. We were in the command post and partway through the lesson when a message came through the SAC command network: the president had been shot. Another message a few minutes later: the president was dead. Maj. Humann called off the lesson, commenting that things might get popping there pretty soon. In those hair-trigger days, it was fortunate they did not.

Special Gunner

While I was assigned to the Mather bomb wing I was crewed up with a gunner named Ukelele Seumalo. He got permanent change of station orders to Barksdale AFB, Louisiana, went for a while, didn't like it, announced he was going back to Mather, and he did. It seems he was a Samoan prince, had diplomatic immunity, and they would station him pretty much anywhere he liked.

Car Models

In the late 1960s there was a constant stream of high-ranking government officials and politicians passing through Guam. Lt Gen Alvin Gilliam, the division commander, liked to ask them what kind of car they owned and how old it was. He would then comment that his bombers were over ten years old and that's why he needed new ones. That was almost 50 years ago and of course they are still flying.

The John

Above a urinal in the men's room of the EW school at Mather there was a brass plaque reading "At this spot on 18 Aug 71 USAF Chief of Staff Gen. John D. Ryan relieved himself."

CAFI

There is a matter I don't believe has been mentioned in this series that truly puts the "dog" in "crewdog" – the Commander's Annual Facility Inspection, or CAFI. This was originally a higher-headquarters check on the local civil engineering squadron, which was responsible for maintaining the base physical plant and grounds. Since there were few ways that wings and their commanders could compete with each other this rapidly grew into an all-out white-gloves inspection of everything in that category that could possibly be inspected. There was a base-wide orgy of cleaning, painting, polishing, waxing, fixing, trimming, and concealing that went on for weeks before the great event, and since a typical bomb squadron consisted of a couple of enlisted admin people and well over a hundred crewdogs, it fell to the crewdogs to do the dirty work. One very talented young pilot of my acquaintance came upon a lieutenant colonel down on his hands and knees scrubbing a latrine and decided then and there to ditch the military at his first opportunity. He later had a great career with the airlines, observing that they had a totally different value system that actually rewarded flying for the sake of flying. I wonder how many others made that choice.

Snake on the Plane

On a combat mission out of Thailand one day our gunner called up from the tail and announced "I've got a snake back here." Someone suggested he depressurize and asphyxiate the snake, which he tried, but it didn't seem to bother the snake much. They uneasily coexisted until we got down a few hours later. It was a good thing the gunner didn't have snake phobia, because I can't think of a worse place to be cooped up with one.

The Mather Mishap

In March of 1961 a Mather B-52 loaded with nuclear weapons ran out of fuel and crashed near Yuba City, California, about 50 miles north of Mather. It is customary to have two accident reports for such incidents, one releasable to the public, the other containing inconvenient truths that are forever classified in the interest of obtaining candor from those involved. There is no dispute over the basic facts: it was a long-duration airborne alert mission that encountered runaway cabin heat about an hour after takeoff, about 12 hours later the heat shattered the inner pane of a cockpit windshield. At first the crew depressurized to keep from blowing out the outer pane. Later they descended to 12,000 feet far out over the Pacific, which increased fuel consumption to the point they couldn't make it back to Mather. A tanker was launched but the B-52 ran out of gas before they could hook up, the crew bailed out successfully, and the aircraft crashed into the Sutter Buttes with nuclear bombs still on board with no detonation of any kind.

There are several accounts of this incident in circulation with varying details, but a persistent question is why the crew continued the flight under those conditions. The crash was a little before my time, but some of the people who were directly involved were still at Mather when I arrived and this is the story I got. There had been a similar incident involving runaway cabin heat a few weeks before and the crew wisely elected to abort, but the upper echelons of the wing staff disagreed. The next day people in the hall could hear the deputy commander for operations screaming at the hapless aircraft commander for his bad judgment in throwing away all that flying time. So when the same thing happened a couple of weeks later, the A/C told the crew "Sorry, guys, but we have to take it."

Runaway cabin heat is bad news, especially on a long-duration mission, and I can imagine the physiological state they were in – dehydrated, wrung out, mentally and physically exhausted, and probably not caring about anything much but getting it over with. It was also starting to look not so good for the wing staff that put them up there in the first place. A tanker launched too late and the rest is history. It is a fact that the wing commander, Col. Frank Amend, was summarily dismissed and never heard from again. For anyone who has been subjected to the inexorable pressure in SAC to make an on-time takeoff and fly out your time, this account rings true.

One small footnote: the official report notes without comment that the gunner bailed out the navigator hatch. In the B-52 F-model the gunner is in the tail without an ejection seat; he pulls a handle, the whole tail end falls off and he simply steps out into space. On this flight, when they found themselves in a big glider and gave the bailout order, the tail wouldn't separate and the gunner had to come forward along a long and narrow catwalk through the wheel wells and bomb bay. The book says this takes about five minutes, but he made it in two and was ready to dive out the open navigator's hatch when the radar navigator stopped him and had him put a chute on first.

Victor

In 1978, apparently by lottery, a number of crewdogs were invited to attend a super-secret briefing on base. This turned out to feature Lt Viktor Belenko, the Soviet pilot who made off with a Foxbat interceptor and defected to the West. His talk was mostly about the incredible dysfunction within the Soviet Union, the lies they grew up with and accepted about how terrible life was in the United States, and the hard time he had believing his own eyes when he got to see the U.S. for himself. He thought it was a setup when they showed him a typical American grocery store – he didn't think all grocery stores could possibly be stocked up and accessible to everyone like that. At a military recreational marina he saw two boats tied up, one very nice, the other – his words – shitty. The nice one belonged to a sergeant, the other to a general. Communism was supposed to have the classless society, but their officers were practically royalty while enlisted were swine, and he was incredulous our system worked the way it did. There was still some risk that the KGB would kidnap him, or worse, hence the intense secrecy. He said that if any of us happened to recognize him on the street it was okay, but don't make a big deal of it. Some years later I saw him at the Reno Air Races, where he was guest of honor and was paraded past the stands before tens of thousands of people. Amazing how things change.

Staying at Castle

In my first ten years in the Air Force, with multiple training schools and temporary duty assignments to the Vietnam War, I never spent more than a year in any one place. About the time the war ended I got assigned as a flight instructor at Castle, where every prospective crewdog went to get checked out in the B-52, and I haven't moved since. With this stability I was able to get married, finish up my undergraduate and graduate degrees, and acquire some new skills. Lt Col Toki Endo, who has written a great Crewdog story about his bailout experience, picked me for a job in Training Devices, and that turned out to be another life-changing event. Flight simulators were undergoing a massive change from mechanical monsters to sophisticated computer-driven engineering marvels, and I was working closely with various aerospace contractors in the process. Along the way I learned enough about software programming and engineering management to get qualifications in both, and on the side I established a lengthy part-time college teaching career in those subjects.

After I retired from active duty in 1983 I came back in Civil Service to almost the same job. There was a peculiar SAC-level detachment at Castle devoted to command-wide B-52 simulator support, including acquisition, testing, maintenance assistance, and developing software upgrades. We had about 60 people, evenly divided between officers, enlisted, and civilians, and I ended up as director of the software support center. I think I was the only civilian on base who was supervising officers, and I wrote far more Officer Effectiveness Reports and Airman Performance Reports than I ever had as a crewdog. The base closed in 1995, and after 35 years in the Air Force in close association with the B-52 I finally cut the cord. Rather remarkably, in a business where there is a lot of age discrimination, I found I was actually employable as a software engineer and got a great job with a start-up company in the booming network industry. So there is life after Uncle Sam, but I shall be forever thankful to that construction guy who by chance walked through the door those many years ago.

No Hablo!

I outlasted SAC and I outlasted Castle AFB but I haven't outlasted the B-52. Some strange things happened at Castle around 1994-1995 as the base prepared to close. Security had always been very tight but some civilian businesses were allowed to move in, and the runway was open to civilian traffic as long as they paid a landing fee. One night a Cessna landed without tower clearance and was immediately surrounded by the Air Police. Two Hispanics were on board who had mistaken Castle for the nearby Merced municipal airport, and after some discussion it was decided that as long as they paid the landing fee they were free to go. The Hispanics then had a conversation in Spanish, assuming that no one would understand them, which is a very bad assumption in Central California. They were talking about drugs, the cops caught it, the airplane was impounded and off they went to jail.

Neighbors

A Russian entrepreneur set up a blimp manufacturing plant in the big B-52 maintenance hangar and from time to time we would see Russian blimps shooting landings along with the usual B-52 and KC-135 traffic. Advertising blimps would occasionally come in for routine maintenance, which I guess is hard to come by in the blimp business. I interviewed there for a possible computer programming job, but what they needed was someone who could convert some FORTRAN programs from Cyrillic to English, a skill I had somehow never acquired. For some years I lived a few houses down the street from a retired brigadier general who had been wing commander at Castle, and in between lived a Ukrainian engineer who worked for the blimp company. After all those years of manning the big nuclear guns we had aimed at each other, all I can say is, it's a funny world.

Patching History

Tommy Towery

This is not a long story, but someday it may answer a lot of questions for historians.

They may be any shape, size, and any color. They range from crude to beautifully artistic. They are starting to show up in flea markets, antique stores, and online auction sites, and someday someone will try to figure out exactly what they are and where they came from. In many people's collection of mementos of their military lives are the one or two odd patches which do not seem to belong with the other "official" patches which once adorned flight suits and helmet bags. They just don't seem to fit the chronology of unit or aircraft assignments. There is no real classification I can put on them except to call them "patches." Many are used as illustrations in this book.

Should someone, someday, decide to do a graduate thesis on the patches worn by USAF personnel, he or she will have their work cut out for them. Give up now, there's no way!

I know B-52 Crewdogs are not the only ones who have them, and I am just as sure someday someone will spend a lot of time trying to research the historical background of the meaning of many of them. I already see it happening in some areas. The truth is researchers will be wasting their time unless they understand one fundamental truth about B-52 related patches during Arc Light days. All it took to have a custom patch made was someone with an idea (however odd), a pencil and paper (even a used bar napkin), a little money (not much), and a village shop (many to choose from) in Okinawa or Thailand and a little time (sometimes less than a day) and voila, a new patch was born. It was kind of like a coward's way of getting a tattoo, but a lot less permanent and painful.

Some designs were widely popular and duplicated and purchased by many, but no one really made any profits to speak of. I could never begin to explain all of them, but there are a few in my own collection which I know about. To help out future historians I will explain a few of them.

For those who served, the term "Golden Flow" is remembered as the nickname of the urine test given to determine if the donor had illegal drugs in his system – primarily marijuana. At random times service personnel were subjected to no-notice drug tests. They were handed a container and asked to fill it up. Often the monitor of these tests had to actually watch the urine flow from the participant's body into the cup. This simple procedure was enough for someone to decide to commemorate the action by getting a patch made.

One of my patches (hiding somewhere in my storage boxes) is a small olive drab rectangular patch depicting a red arrow with a wiggly shaft. The words "Straight Arrow" is embroidered beneath the arrow.

Mine came from Thailand, where the term "Straight Arrow" meant that an airman didn't fool around on his wife or girlfriend with the beautiful Thai girls in the clubs. You can put your own slant on why the arrow really isn't straight. And while we are on the subject of beauty, unlike the movie "10" where Bo Derek was so beautiful she was rated a ten on a scale from one to ten, in the U-Tapao community "Number 1" was the top of the scale. Unfortunately some of the prettiest Thai ladies were actually Kathoeys – males who are female impersonators. "Ah, Number One good deal." In Okinawa the term for number one is "Ichiban."

Another patch made in local shops during the Arc Light period which was very popular was the "Southeast Asia War Games Participant" patch, which was a take-off from the Olympic Games. There are many different versions of these still around, some of which are being reproduced today.

Custom nametags were also very popular, whether they had your real name, your nickname, or someone else's name embroideried on them. Besides the Yossarian one above, another famous nametag was that of "Sam Uplink." I remember hearing the story of the Haircut Colonel telling a pilot to get his EW "Sam Uplink" to get a haircut, and he was going to check later because he had written down his name in his little book. It was obvious the Colonel was unaware that "SAM" was short for "Surface to Air Missile" and "Uplink" was the guidance signal which directed a SAM after it had been launched at an aircraft.

This brings me to the true story of the pride of my unique patch collection. My crew Carswell R-04 (consisting of Pilot – Denny "Scruggo" Scruggs, co-pilot Tom Adcox, R/N David "Mad Dog" Kerr, Navigator Rich Hayes, EW, me and Gunner Doug Bunch) was ferrying a plane from Andersen AFB, Guam, to U-Tapao during one of our deployments. A high ranking Andy staff officer had promised to send some fresh steaks to his buddy at U-T and selected us as the delivery boys. He had them strapped down in the plane's wheel well area to keep them frozen during the flight. Several hours after takeoff we lost pressurization in our plane and our first thought was to divert into Clark AFB in the Philippines. When we notified the command post of our intentions we were told we could not do that because of the political situation, and we were to drop down to an altitude of 8,000 feet where we did not need oxygen and continue on to U-T. At that altitude the wheel well area failed to keep the steaks frozen, and by the time we landed they had all thawed out so the recipient ended up having a large bar-b-q for a lot of people which he had not planned on having.

The Flying Safety Officer was upset because we continued to fly a malfunctioning B-52 to U-Tapao rather than aborting the mission and returning to Andersen, which was closer to where the problem happened. He threatened to write up the crew with a safety violation and was serious about doing so until he was stopped by a higher ranking staff officer. We never knew if it was the "steak man" or not, but somehow we were off the hook. We were shocked to find out we had been designated the "Safety Crew of the Month" for U-Tapao that month – even making the list in the Air Force Magazine. We always felt it was revenge by the Flying Safety Officer because we made him look bad to the staff. He gave us the honor as a way to get even with us because our reward it turned out was to get to brief all the crews (several at a time) on our inflight emergency at the Weekly Flying Safety Meeting for the entire month. This required us to stay around the base and work on our days off for those days, so in our eyes we were screwed.

We went down to one of the shops at Sattahip and had patches made for the whole crew. The patch was a design using a condom with a safety pin through the end of it, and the words "Scruggo's Safety Screw" around the edges. For the historian's sake to translate the hieroglyphics of the patch, "Scruggo" was the nickname of our pilot, the safety pin symbolized our safety award, and the condom depicted the fact we were screwed.

For every patch on every flight suit, there is a story to go with it – you just have to dig to find it before it is lost forever.

TV screen capture of landing at Duxford.

Delivering a "British" BUFF

Steve Winkle

The Imperial War Museum Duxford, located near Duxford in Cambridgeshire, England is Britain's largest aviation museum and a branch of the Imperial War Museum. In 1983, the USAF agreed to place a B-52D on permanent public display at the American Air Museum located there. The aircraft chosen for delivery to England was B-52D 60-689.

The bomber came from the 7th Bomb Wing, Carswell AFB, Texas, and the crewmembers chosen to deliver it were Lt Col Jim Nerger, Aircraft Commander; Lt Col Wally Carpeaux, co-pilot; Maj Steve Winkle, Radar Navigator, and Capt Wes Hewett, Navigator. Maj Joe Schwab would also be aboard as a relief pilot and celestial observer.

Mission planning was straightforward and routine. The route was essentially a Great Circle route taking off from Carswell and landing at RAF Brize-Norton located 12 nautical miles west of Oxford. The total route would be just over 4,100 NM and take around 10 hours flying time. The only hitch expected in the entire mission would be encountered after arriving at Brize-Norton and then taking the aircraft on to Duxford. That's because Duxford's runway is only 4,000 feet long - making it the shortest runway on which a B-52 had ever attempted to land.

This fact required the crew to fly a practice "pilot proficiency" mission at Carswell to see if, in fact, the aircraft could even attempt landing in less than 4,000 feet. At this point I should mention that Lt Col Nerger had somewhere north of 10,000 hours in the B-52 and was firmly convinced such a landing was not only possible, but, for him, a fairly simple matter.

We did fly a mission in the Carswell pattern with numerous touch and go landings followed by a full-stop with a fuel load approximating what Nerger and Carpeaux calculated we'd have at Duxford. The final "practice" landing demonstrated to the 7th BW DO that a 4,000 foot runway was no problem and the mission was approved.

The route of flight was northeast from Carswell, out over the Canadian Maritime provinces direct to the United Kingdom. The first "challenge" was the 2,000 NM leg overwater with no navigation aids. Not a problem; I was an Instructor Radar Navigator with around 3,000 hours in the BUFF and Wes was an experienced Navigator with a bunch of hours as well. Here was a chance to do a real nighttime celestial overwater mission!

At this point, I should also mention that, on a regular B-52 crew, celestial observations were performed by the Electronic Warfare Officer (EW) using the computed information from the Navigator and observing and recording the actual, observed readings from the sextant. One thing no one on the delivery crew knew was what the EW did to "preflight" the sextant. This would later prove to be a problem.

The flight from Carswell to "feet wet" over the Atlantic was routine. It had grown dark with sunset and I got a "final fix" just before leaving the coast behind and heading out over the ocean. The B-52D had no inertial navigation system, so we only had the radar and celestial observations to guide us the 2,000 miles overwater.

Wes had pre-computed the nav's celestial information and I checked and concurred with his computations. The observation schedule was, if I remember correctly, 12, 8, and 4 minutes prior to "fix" time and there were to be three celestial fixes total before we were within radar range (approximately 200 nm) of land.

Joe got the sextant out of its case and inserted it though the sextant port in the roof of the plane. He then set the pre-computed angle and azimuth given him by Wes to find the star Wes planned to use for the first observation. Normally, a navigator plans to use three stars located approximately 120 degrees apart to observe for his navigation "fix."

When Joe turned to the azimuth Wes gave him, there was no star in the field of view! Oops. Something was wrong. However, Joe did find a bright star 15 degrees to the right of the azimuth provided by the navigator. We went ahead and used this star for the first observation. Unfortunately, the subsequent two stars were also 15 degrees off from the expected azimuth. And when Wes plotted his first fix, it put us north of track. He gave the pilots a 15-degree heading change to the right to get back on track. Do you notice any commonality at this point? If you did, good; because none of us aboard the flight did.

The following two fixes again showed us north of track requiring a right turn. We figured the Jet Stream was hitting us on the beam and pushing us to the left of where we wanted to be.

About 45 minutes before our expected landfall, I went out to the long range mode on the radar hoping to pick up the coast of Ireland so we could get a radar fix and verify our position. At roughly 200 nm, I began to detect the Irish coast. But it didn't look like I expected it to look.

What had happened was that the azimuth ring that is part of the sextant mount had shifted 15 degrees and that shift, in turn, made the azimuth the celestial observer used off by that same 15 degrees. As part of a real EW's sextant preflight, the EW would have checked that the alignment of the sextant azimuth ring was accurate. Since we didn't have a "real" EW aboard, our observer didn't know to do that. In going back and re-plotting the mission using the corrected azimuth, Joe's observations and Wes's computations would have put us right on track for the entire celestial portion of the route.

When I was able to get a confirmed range and bearing on the radar off the Irish Coast, I realized we were seriously south of track and told the pilot to come 45 degrees left. It wasn't much longer before we received a radio inquiry from the British Air Sector controller asking why we were off course. We acknowledged we were south of track but were already correcting back to course.

With the radar now verifying our position and getting under the radar control of British Air Route Controllers, we were able to complete the flight into RAF Brize-Norton.

Even with the difficulties encountered, that was the easy part of the trip. Now, the pilots had to familiarize themselves with the routing and visual aids they would use to bring the delivery to a successful ending at Duxford. To do this, they were each taken up by the RAF in a RAF aircraft and flown from Brize-Norton to Duxford. This was a short flight of only 66 nm.

Jim planned to do several low passes over the runway to ensure he had a good feel for his final approach. The Imperial Air Museum is just to the west of the town of Duxford. Between the town and the airfield is the M11, a major freeway going north out of London - just 25 nm to the south. Because of the intense media interest in this delivery, the officials closed the M11 during the period the aircraft was in the pattern.

The minimum crew for a B-52 is a pilot, a co-pilot, and a navigator. In addition to Lt Col Nerger, Lt Col Carpeaux, and myself, we also had a RAF flying officer aboard. Unfortunately, these many years after the event, I no longer remember his name.

The flight was routine and kept to fairly low altitudes between Brize-Norton and the airfield. Jim and Wally performed several low passes much to the delight of the crowds below and then lined up for the final approach and landing.

Everything was routine and we touched down on the target at the end of the runway, Wally deployed the drag chute, and both he and Jim got on the brakes. The aircraft came to a full stop with 2,000 feet of runway remaining! This became, and to the best of my knowledge, remains the shortest landing for a B-52 in history.

Today 60-689 sits inside the glass enclosure of the American Air Museum along with several vintage USAF aircraft from WW-II, Vietnam, and the Gulf War, and remains a favorite of many visitors.

The Year Bomb Comp Wasn't!

Derek Detjen

It was the late summer of 1967, and our crew had just returned from a long, all night mission landing early in the morning at Andersen AFB, Guam. We were Columbus AFB, Mississippi's crew E-13 and all the officers on the crew (Maj Joe Steele - Pilot, Capt Fred Luciano - co-pilot, Maj Cliff Brown – Radar Navigator, Capt Dave Shell – Navigator, and myself – Electronic Warfare Officer) decided to go to the Officers' Club for dinner during "happy hour" and headed for the bar before eating.

There was a group of fellow aviators sitting close to us, and their English accents readily identified them as non-SAC aircrew personnel. Introductions all around soon followed, and we learned they were in fact a British RAF Vulcan bomber crew. They had been chosen by the RAF as their top rated crew and had been sent to the U.S. to participate in our annual bombing competition. Unfortunately the big wigs of the Strategic Air Command in Omaha had decided at the last minute to cancel the competition, possibly due to our increased presence in Southeast Asia.

The RAF decided that the crew's consolation prize would be to continue their planned round-the-world flight plan to get back to the U.K. This route consisted of visiting a west coast U.S. base, then on to Guam and Singapore before returning home. The evening meal sped by rapidly, due to our lively discussion, until suddenly everyone became silent. Our attention was drawn to their bombardier (Nav) who had fallen into a deep sleep, with his nose buried in his steak. His snooze was quickly blamed on the crew's lack of sleep the past few days.

As an aside, their crew also informed us that unlike the bomber crews in SAC, in the RAF, one upgraded from bombardier to navigator (Nav-Plotter), as that position was deemed more complicated. This information was much to the delight of our own Dave Shell! Before we departed the club, the Brits invited us out to the runway the following morning to watch their scheduled 0830L takeoff time and promised us a "good show." Happily we had already been scheduled for taxi-crew duty in the morning, so we quickly accepted their offer and were already on the grass near runway 06R as the big delta-winged bomber taxied into position.

The scream of their engines punctuated the quiet morning as they accelerated down the downhill portion of the runway (about 2,500 feet) and just as they reached the bottom, they rotated and pulled the nose literally 90 degrees up, standing on their tail as they quickly vanished from sight. We had just returned from a short stay at Kadena AFB, Okinawa, and one morning held short of the active to let a SR-71 pass by and make an almost identical takeoff (except for the little H-34 helicopter that always followed them with their fire-fighting sled in tow). The SR-71 disappeared just as quickly, with a similar 90 degrees nose-up trajectory.

Capt. Detjen (right) and Gen. John P. McConnell (U.S. Air Force photo)

Later that year, in October, we were visited by Gen Paul McConnell (then CINCSAC), who presented the entire crew (including our gunner, MSgt Red Baird), with our fifth Air Medals in recognition of being the first B-52D crew to reach the 100 mission plateau. Although we were justly proud of our accomplishment at the time, it was a mark easily surpassed by many other crews who completed multiple Arc Light tours from 1968 on.

Soon after completing our third tour the following year, I was selected to join Castle AFB's Replacement Training Unit (RTU), and for the next four years we trained over 600 G and H-model crews into the older D-model before their departures to Guam. RTU's instructor force averaged three tours and some 160+ missions each. All of us enjoyed the relaxed training environment, just sharing information and preparing them for their upcoming Arc Light tours.

The BUFF Goes on Holiday

Paul Paris

In October, 1981, the 7th Bomb Wing at Carswell AFB, Texas, was tasked to deploy three B-52Ds to support a NATO exercise. Their destination was Fairford Royal Air Force Base, located 70 miles west of London. The mission's significance was underscored by the fact it would be the first time in history a B-52 would touch down at Fairford. Just before sundown on 19 October, I was the IP on the third aircraft of the three-ship cell taking off in one minute intervals. I was flying with Crew S-01, the most experienced crew in the squadron, to augment the other two pilots. I had arranged for my wife and son to watch the departure from the Supervisor of Flying (SOF) vehicle and was excited for them to be there, since my five year old son had never seen his dad take off before, and it was my wife's birthday.

Things went as planned when we assumed our one-mile spacing after leveling off at cruising altitude. An hour after takeoff we had no problems when we picked up a Carswell tanker over Oklahoma, but the second refueling off the coast of Nova Scotia, with a Loring AFB, Maine, tanker did not go as well. With bad weather, we had difficulty rendezvousing with the tanker, and by the time we eventually hooked up, we didn't have much time remaining on the refueling track. Turbulence made it very difficult to stay connected with the tanker. Capt Dave McCracken, one of the most experienced pilots in the squadron, commanded our aircraft. Prior to refueling I traded places with the co-pilot, Lt J.J. Wright, because I knew getting our gas was going to be critical. After two disconnects, the time on the refueling track was rapidly vanishing, and thoughts of diverting into Loring AFB, Maine, danced in our heads. No one wanted an unexpected TDY at Loring by the Sea.

When Dave made the third contact, I reached up and switched the refueling system to MANUAL without telling him. This setting overrode the AUTO feature, and trapped the boom toggles. Had we disconnected in the MANUAL mode, we would have taken part of the boom home with us. The procedure was reserved for wartime or emergency situations only, but I considered it an emergency to avoid landing at Loring instead of Jolly Ole' England. Dave was fighting to stay in position with the tanker, and I didn't even tell him what I was doing. He held on for a few minutes before we reached the end of the refueling track. I had already told the navigator to work up a flight plan to Loring, since it didn't appear we were going to get enough fuel to make it to England. However, I guess we got lucky and our aircraft on-loaded just enough fuel to make it to Fairford. We had to cancel our planned high altitude bomb runs over Ireland and therefore landed at Fairford nearly three hours early.

After flying 11.5 hours, we landed around 0630 on a cold, drizzly Saturday - a typical British October morning. After the long trip I was about half asleep until I noticed something I thought was very strange. As we turned off the runway and began our taxi in, I was stunned to see a throng of civilians lining the outside of the barbed wire fence serving as the boundary of Fairford and the rest of the town. The fence was not too far from the taxiway, and I could see their faces as I leaned forward from the IP seat. Old people, young adults, and children had watched us land, and then taxi into the parking area. I didn't see any protest signs or people yelling at us. They stared with expressionless faces, not seeming to even mind the cold rain. They were neither happy nor angry

- they were simply curious. They were up early to see history being made. I wondered what they were thinking. Seeing all those people was an incredible feeling for me. How would they recount this event to their families? Would they put this experience in writing in the form of a memoir? Would they embellish the story to make it more exciting? I felt very special as I watched them watch us. I was a legend in my own mind.

Three days later, I was tasked to give a tour of the B-52 to a group of British cadets and their commander. The group consisted of 13-15 year olds who were pumped up, to say the least, to see the inside of the legendary American bomber. But their enthusiasm didn't cause them to be unruly and out of control as their American counterparts might have been. Instead, they showed me the utmost respect and were painfully polite, in the usual British manner. They went inside four at a time, and inspected every inch of the forward compartment. I was surprised at how much they already knew about the BUFF; they had obviously done their homework. After each set had toured the inside, we met as a group outside, underneath a wing. They pummeled me with questions for nearly 20 minutes until their commander announced they had to return to school. I answered each question with pride as I observed the excitement on their faces.

After returning to Carswell, I received a sincere "Thank You" letter from their commander who assured me the tour was the highlight of the year for the young boys. The short note certainly affirmed my pride in being a Crewdog.

Résumé

Résumé [rez-oo-mey] - *noun* - brief written account of personal, educational, and professional qualifications and experience.

The Contributing Authors

Peter Bellone

Master Sergeant, USAF, (Ret.) retired from the Air Force and went to work for McDonnell Douglass as a Logistics engineer then with the US Army as a Records Manager. He retired again as a DA civilian in 2011. Presently he seeks contracts for his business - Bell1 Logistics, LLC, and is a published author of *When Time Finally Runs Out* which can be found on Amazon.com as well as barnesandnoble.com.

Pat Branch

Colonel, USAF (Ret.) in 1962, after graduating from college and being re-designated from 2S to 1A draft status, it was difficult to find a job that would support my growing family. I enlisted in the Air Force and immediately went to Officer Training School with the full intention of returning to civilian life as soon as possible. Thirty years and 14 assignments later I retired as a colonel on the 30th anniversary of commissioning. My career can be divided into two halves: Electronic Warfare and Intelligence.

The flight pay of $100 a month was my motivation to pursue a flying career. Pilots and Navigators got the same pay, so I jumped at the chance to be a navigator. Most rated officers will tell anyone who ask that they "always wanted to fly" or that they were motivated by watching World War II fighter aircraft. That's B.S., I had bills to pay and babies to feed.

While in Nav School, I finished top in my class in electronics and the die was cast - I was going to Electronic Warfare School. Over a 12 year period from 1965 to 1976 I would fly almost 5,000 hours from six bases. It took seven years of BUFF time to accumulate 2,000 hours but less than half that long in the RC-135. I also few the B-58 and the B-66C/E for exactly one year and accumulated approximately 350 hours in each aircraft.

Highlights of the flying career included a B-52 Arc Light tour at U-Tapao, six Chrome Dome (24-hour airborne alert) flights near the Greenland BEWS site, and, of course, the infamous SAC ORIs. During Linebacker II (December of 1972) I flew B-66E Electronic Warfare and B-66C ELINT reconnaissance missions. By far the most enjoyable year of flying was the B-58. While the DSO (EW) did not have Pilot's or Navigator's controls, he was crucial to the crew. Pilots went to fly with a three-page checklist. DSOs flew with a nav-bag full of tech orders, maps, and slip sticks. The RC-135 missions flew from Offutt AFB, NE; Alaska; Kadena, Japan; Mildenhall, UK; and Greece.

In 1976, I was pulled up to the SAC staff and took over a small division writing requirements documents and concept of operation studies for future reconnaissance systems. My days as a crewdog were over. From that assignment to my final retirement, I was deep into the world of intelligence. In addition to Hq SAC, my staff assignments included ESC, USAFE, and the Pentagon. With command assignments at RAF Mildenhall (RC-135), Hahn AB, Germany, (Director of Metro Tango - U-2 ground station) and USAFE 7450 TIW/CV. My final forward deployment was immediately after the Gulf War as the Provide Comfort, Combined Forces Command, Chief of Intelligence. I returned to the Pentagon as the DIA Director of Intelligence Collection.

After retiring from the Air Force in May 1993, I worked as a defense contractor embedded in the Under Secretary of Defense staff with responsibility to integrate the Services intelligence collection and dissemination capabilities through a program know as Distributed Common Ground Systems DCGS,

This diversity of assignments may seem rather bizarre to those now on active duty. Back then, changing positions on a two-year average over a 30 year career, you got to do a lot of things. Catherine and I have seen a lot of changes in our 50+ years together. We started in the Air Force to solve an employment issue. We enjoyed each new assignment more than the previous assignments and are now enjoying playing with our grandchildren back home in Minnesota.

Michael L. "Mike" Brinkman

Former Capt., USAF (Honourably Discharged), was commissioned in 1972 following four years of ROTC at The University of Alabama. After busting out of T-38s in Undergraduate Pilot Training at Moody AFB, Valdosta, GA, he headed for Mather AFB in Rancho Cordova, CA and Undergraduate Navigator Training. His class was the last to fly in the venerable T-29 twin prop trainer. Graduating in 1974, Mike went through Bomb/Nav. training at Mather and then on to Combat Crew Training School for B-52's at Castle AFB, Atwater, CA. Once again Mike was in "a last": The last class that flew in the B-52F as non-integrated crewmembers.

From Castle it was on to Robins AFB, GA. With all the integrated crews arriving from Castle already qualified in the B-52G model, Mike ended up being a spare navigator and the Bomb Wing Civil Engineering Liaison for one and a half years. He volunteered to PCS to Anderson AFB, Guam. His wife Kathy and Mike loved Guam. Both daughters were born there. After five years in the "Dog" model BUFF, AFMPC tried to figure what would be the most contrary to a lush, mountainous, tropical island - how about Grand Forks, ND? In1981, Mike started off in the H Model until the 319th Bomb Wing switched over to the B-52G and Air Launched Cruise Missiles. At that point, he went on the wing staff as the Emergency War Order (EWO) Training Officer and a Coded Switch Custodian.

Mike and Kathy left the service in early 1985, and the family returned to Mike's home of record: Huntsville, AL. He had earned a Master of Aviation Management degree from Embry-Riddle Aeronautical University while at Grand Forks. This came in handy during his job search. He was hired by a NASA contractor at Marshall Space Flight Center to work on the Space Shuttle Spacelab payload (experiments) integration and operations program. After eight years in this line of work, Mike received the opportunity to explore other lines of work including selling annuities, working in retail, and small manufacturing. After five years, he was asked to come back to payload integration and operations, this time for the International Space Station. This was followed by an Army Aviation contract, and, finally, the National Missile Defense program.

Then an amazing thing happened: God called Mike into pastoral ministry in 2001. Beginning in June 2002, he began serving a church while attending seminary at Emory University, Atlanta, GA. After three years, Mike graduated with a Master of Divinity (MDiv) degree.

In 2008, he was ordained as an Elder in The United Methodist Church. Having served a number of churches in North Alabama, Mike retired from pastoral ministry in June of 2015. Currently he and Kathy live in Oneonta, AL. One daughter is married to a USAF Lt. Colonel, and the other is married to a government communications technician and TSgt. in the Alabama Air Guard. Air Force blue runs in all their veins, including those of the four grandchildren.

Christopher Buckley

Lt Col, USAF is an active duty B-52 instructor/evaluator Radar Navigator. He is a master navigator with over 2,500 hours in 45 different aircraft and is a graduate of the USAF Test Pilot School. He has flown the B-52H with the 23d Bomb Squadron (Minot AFB) and the 419th Flight Test Squadron (Edwards AFB). His story does not have the endorsement of the United States Air Force or the Department of Defense.

Ken Charpie

Lt Col, USAF (Ret.) earned a BS degree in Chemistry in 1978 from the US Air Force Academy. He stayed in the Astronautic Engineering Department doing research until time to head to UPT at Vance AFB, OK in November 1978. After washing out of UPT, and a bit of a delay awaiting orders, he attended UNT at Mather AFB, CA. UNT was followed by Advanced Navigator Training, Water Survival, Nav-Bombardier Training, and finally B-52 CCTS at Castle AFB, CA. His first B-52 assignment was in the 5th Bomb Wing, 23rd Bomb Squadron, Minot AFB, ND, in the B-52H. He started out there as a navigator on Crew E-23, and upgraded to Instructor Navigator, then was assigned to Crew S-02 as a Stan Eval nav. Following his stint in Stan Eval, he was sent again to Castle for B-52 Radar Navigator Upgrade training, with return to Crew R-35 at Minot. Yet one more trip to Castle saw him complete Central Flight Instructor Course, and become an Instructor Radar Navigator.

After five years at Minot, the Air Force sent him to the Air Force Institute of Technology for a MS degree in Operations Research, majoring in Strategic and Tactical Sciences, and graduating in 1987. To benefit the Air Force with this degree, he was assigned to the Air Force Operational Test and Evaluation Center at Kirtland AFB, NM. He primarily worked on the B-1B and B-1B Simulator System operational testing, but also got involved in a number of smaller test programs, such as Short Range Attack Missile II, Advanced Cruise Missile, and KC-135 Wingtip Air Refueling Pods. He completed his AFOTEC assignment early in 1991, but had to wait for the B-52 CCTS instructors to return to Castle AFB from Operation Desert Storm.

His B-52 Requalification training was completed in the B-52G since that was all Castle was flying at that time. After completing the requalification, he reported to KI Sawyer AFB, MI, and the 410th Bomb Wing, 644th Bomb Squadron to join Crew E-75 as an instructor radar navigator, and as the flight commander, just in time to see B-52s taken off alert. During three years at KI Sawyer, he also moved into the wing staff as it was reorganized into the Air Combat Command objective wing structure, becoming part of the 410th Operations Support Squadron, first as the EWO Certification Officer, and ultimately moving into the Operations Flight Commander position.

As KI Sawyer AFB closed, he received a new assignment to Air Combat Command's 31st Test Squadron at Edwards AFB to be the Branch Chief of the Data Management Branch on the B-2A bomber's operational test program. He also became the only member of the OT&E team qualified to serve as a test director in the control room during test missions. He accomplished that duty for the remainder of the operational test missions, and also for some of the developmental test missions. As test director, he had to make real time decisions affecting flight safety not only for the B-2 being tested, but also for all support aircraft for the mission, chase aircraft, and tankers. One notable mission saw high winds arise during the test, necessitating immediate landing. Due to different cross wind capabilities of the B-2, KC-135, and F-16s involved, he had to direct planes to land at the Edwards dry lake bed, Palmdale, CA, and Vandenberg AFB, respectively.

Upon conclusion of the test program in 1997, he was assigned to HQ Air Combat Command, Directorate of Requirements, ACC Systems Office at Wright-Patterson AFB, OH as a liaison officer to the B-2 System Program Office. In the B-2 SPO, he constantly represented the voice of the user in meetings with the B-2 SPO Director and other staff personnel. In 2001, he retired from active duty with over 23 years service, and approximately 2,100 hours in the B-52H, and 90 in the B-52G.

After retirement, he spent 12 years working as a support contractor in the B-2 SPO, and retired from that. He currently teaches two classes a semester at Wright State University, teaching students and faculty there the fine art of scuba diving. In addition, he has started "righting the wrong" committed when he washed out of UPT and is currently working on his private pilot license at Greene County Airport in Xenia, OH.

Bob Davis

Lt Col, USAF (Ret.) began his career in 1955, shortly after finishing high school in Mineral, VA. He got out in 1959 but missed a certain two-striper at Waco so returned to HQ 12th AF. He re-upped for her (Sally) and a promise of OCS. Sally & Bob married in 1960 and headed for Lackland. In April, 1961, Lt. Davis moved to Harlingen for navigator training, followed by school at Mather AFB, CA for EWO training. Then it was B-52 training at Castle and on to Seymour Johnson AFB, NC as an Electronic Warfare Officer on "BUFFs." He PCS'ed to Westover in 1967 and flew lots of missions in SEA. "...about 140 missions out of Andersen, Kadena and U-Tapao. Bob went to Eielson AFB, AK in 1969 in RC-135S "Cobra Ball." Maj Davis was selected for ACSC in 1972. Career broadening meant personnel school and assignment to Sheppard -- and a tour as Club Officer! Bob was more than glad to return to the cockpit two years later. There was a year in Korea in F-4's, then Bob selected Eglin's Tac Air Warfare Center, "doing all kinds of unmanned systems things." Retired Dec 31, 1981 as a Lt Col w/DFC, 12 AM, MSM, Airman's Medal, plus others. After retirement, Bob earned his MBA and ended up close to the defense industry until re-retiring in 1999. After re-retirement Bob worked for three years as a consultant. He and Sally play golf a couple times a week and are active in civic and church work. He also is currently working as a volunteer with the Fort Walton Beach Sharing and Caring office and with the USO Center at the Destin-FWB airport -- and something about being a "Noles" fan? And, oh yes, I have a son and daughter and nine grandchildren.

Greg Davis

Lt Col, USAF (Ret.) came to the B-52 as a Captain after having been a T-38 IP/FAIP, followed by a tour at the U.S. Air Force Test Pilot School (USAFTPS). Greg had no knowledge of what a SACUMCISM was at the time he arrived in the B-52G. He had one assignment in the B-52G at Loring and enjoyed the assignment with his crew. After two years as a crewdog, He went to CFIC in one of the last classes to get to do the WIFF. He came back to Loring and immediately moved up to be the Chief of Training.

Greg had fun sending B-52s on out and back missions with munitions for training. (This actually drove some people at HQ nuts because after calling them for permission every other week for a month, they told him don't call back and don't screw it up.) Because it was 15 hours to fly to the Utah Test and Training Range (UTTR) and back from Maine, Greg started sending airplanes every other week to Castle and Fairchild to get two missions done instead of one for just a few more flying hours.

For the last two years of operation, Greg had a great staff assist him in keeping the B-52 on top of all the other ACC bomb wings in terms of bomb scores. The rest of the ACC bomb community could not wait to have Loring close due to getting their noses rubbed into the bomb scores. The diversity of the training program was key to making the crews work even on the back yard low level route. One radar nav said he could no longer fly it in his sleep.

Derek H. "Detch" Detjen

Major, USAF (Ret.) grew up in New York City and in the Cincinnati area. As an Aerospace Engineer student at the U. of Cincy, he flew in one of the first KC-135As during its cold weather testing in Alaska. In late 1960, he entered navigator training at Harlingen, TX, finishing as a Distinguished Graduate. At Keesler AFB, MS he met his wife Betty and graduating as an electronic warfare officer.

Assigned to Turner AFB, GA he participated in the first six-month Arc Light tour to Guam and returned again from Columbus AFB, MS in both 1967 and '68. In late 1967, his crew E-13 became the first to complete 100 strike missions. Their 5th Air Medals were presented to them by the then CINCSAC, General Paul McConnell at Guam.

Four years at Castle AFB, CA followed, training over 600 B-52G/H-model crews into the D-model, as a member of Castle's Replacement Training Unit (RTU). As a member of 1CEVG, he then returned to Guam as Ops Officer of Det 24 "Milky" training their now permanent B-52 Wing.

His last five years were spent at Barksdale AFB, LA in charge of B-52 and KC-135 EWO study. He was selected to go to RAF Fairford, UK for the 1982 NATO exercise. His decorations included a DFC, a Meritorious Service Medal, eight Air Medals and several lesser awards.

After active service, he worked on the Trident submarine at NSB Kings Bay, GA for four years, attained his MPA at Valdosta State U., teaching in their on-base program before a final ten years as director of the Management and Marketing majors at Aiken Tech College in the Augusta, GA area. As a lifelong golf devotee of the Masters, he's attended over 50 tournaments, living just 10 minutes from the course.

Russell "Duff" Duffner

Lt Col, USAF (Ret.) is a B52 pilot with over 3,000 hours. Commissioned 1962 (OTS), Chanute Maintenance Officer School '62-'66; Moody AFB '67 Pilot training, (Bailed out of T-32 after a mid-air collission). Assigned to Ramey AFB, Puerto Rico, '67-'70 then Kincheloe AFB from '70-'73, where he was on the Fairchild Trophy winning Bomb/Comp Crew. TDY to Andersen AFB, Guam, and U-Tapao RTAFB, Thailand during the 11 Eleven Days of Christmas (Linebacker II) combat. His crew was known as Bunch 10.

From '73-'76 he was assigned to the ROTC unit at NC A&T State University. Next he moved to Wurtsmith AFB from '76-'79 and then Offutt from '79-'81 service with the SAC IG. His final assignment was LoringAFB from '81-'84.

Flew B52 C, D, E, F, G & H's. He is the recipient of the Distinguished Flying Cross, Meritorious Service Award with 2 Oak Leaf Clusters, Air Medal with 1 Oak Leaf Cluster, Outstanding Unit Award with 2 Oak Leaf Clusters, Vietnam Service Medal with 1 Oak Leaf Cluster.

Retired in 1984 and lives in the mountains of North Carolina in Mills River, near Asheville.

W. Scott Freeman III

Lt Col, USAF (Ret.) was commissioned through OTS in June, 1969. He attended Undergraduate Navigator Training at Mather AFB, CA with follow-on training as an Electronic Warfare Officer (EWO). For his first operational assignment, he served as an Electronic Warfare Officer on B-52D aircraft at March AFB, CA. During that time, he flew 262 combat missions in Vietnam and accumulated almost 1,100 combat hours. Following Vietnam, he transferred his operational expertise to the 55th Strategic Reconnaissance Wing at Offutt AFB, NE, where he flew RC-135 V/W/M model aircraft as a Raven, where he quickly upgraded to instructor and flight examiner.

In September 1977, he left Offutt AFB and went back to Mather AFB, CA and the Electronic Warfare Training School as the Reconnaissance Flight Commander and eventually to the 323rd Flying Training Wing as the Chief, Electronic Warfare Curriculum Development Branch. In September 1981, he was reassigned to Headquarters Air Training Command, Randolph AFB, TX as an Air Operations Staff Officer. He assumed the role as Program Manager for a $10M modification to the Air Force Basic Electronic Combat Simulator and also developed command policies, directives, and syllabi of instruction for all EW training programs.

In May 1984, Lt Col Freeman became a Program Element Monitor (PEM)/Electronic Warfare Program Manager at HQ USAF, Pentagon. In May 1986, Lt Col Freeman moved to the DIA, Pentagon, as the Director of the ELINT Division. He established and directed policy for operational ELINT support and directed all activities as a Duty Director for Intelligence (DDI) in the National Military Intelligence Center during several worldwide crises. In September 1989, Lt Col Freeman moved to the National Reconnaissance Office (NRO) Defense Support Project Office (DSPO) at the Pentagon. He was initially assigned as the Budget Director and developed the DSPO budget for tactical and space airborne reconnaissance support program to Congress. He also directed the $20M MERIT program, an ASD/C31 directed permanent working group, which sponsors potentially new technology insertion programs which augment national programs, increasing the effectiveness of space assets in support of the war-fighting Combatant Commanders.

Lt Col Freeman's final Air Force assignment was on the Intelligence Community (IC) Staff, National SIGINT Committee as the Vice Chairman, SIGINT Overhead Reconnaissance Subcommittee (SORS). He was responsible for the development of Overhead SIGINT requirements for the research, development, and procurement of National SIGINT systems.

He retired in 1993 after 24 ½ years of service in the USAF, and became a defense contractor with TASC in the Washington DC area for 19 years. Scott and his wife Jackie live in South Riding, Virginia and both retired in 2012 to enjoy golf, skeet and recreational shooting, frequent visits to the grandkids in New Hampshire and yearly visits to Hawaii and other destinations.

His awards include Defense Superior Service Medal, Distinguished Flying Cross with V device and 1 oak leaf cluster, Air

Medal with 14 oak leaf clusters, Defense Meritorious Service Medal with 1 oak leaf cluster, Meritorious Service Medal with 2 oak leaf clusters, and numerous other awards.

Russell Greer

MSgt. USAF (Ret.) - I was born in Jacksonville, NC, to Emory W. Greer Jr. and Dorothy Mae Greer. In 1969 my father was transferred with Sears to Tuscaloosa, AL., I was 10 at the time. My father was a licensed private pilot owning an Aeronca Champ and Stinson Voyager as I was growing up. He later built a Rutan Vari-Eze after I left to join the Air Force. I enlisted in the USAF in 1977 as an Automatic Tracking Radar Systems Technician and was stationed at Det. 10, 1 CEVG in Hastings, NE. Cross-trained in 1982 to B-52 Gunner and attended CCTS at Castle AFB, CA. and was then assigned to crew R-06 of the 62nd BMS, 2nd BMW, Barksdale AFB, LA. In 1985 I attended the Central Flight Instructor Course (CFIC) at Castle AFB. In 1985 my crew, E-06, was selected by the 2nd BMW to participate in the annual SAC Bomb-Nav competition. In 1986 I was assigned to crew S-03 as part of the 2nd BMW Stan Eval division. In 1986 our crew was again selected to participate in the SAC Bomb-Nav competition where we finished third in spite of being the only unit still operating the older ASQ-38 Bomb-Nav system. I was deployed twice on TDYs to RAF Fairford to participate in Busy Brewer exercises and would find myself back here in 1991 during Desert Storm. In 1987 I flew as a Bomb Comp monitor with the 42nd BMW at Loring AFB, MA. In 1987 I became the lead gunner in the 2nd BMW Stan Eval division and moved to crew S-01.

In 1988 I was selected by HQ SAC to relocate to the 436 STS (Strategic Training Squadron) at Carswell AFB, TX to become the B-52G Curriculum Development Manager for the command. I also completed B-52H difference training and maintained dual qualification in the G and H models for the rest of my career in SAC. While stationed at the 436 STS I wrote and participated in the shooting of the updated B-52G Strange Filed Disarming course and video as well as the B-52H Strange Field Disarming and B-52G & H Aircrew Servicing courses and videos. I was also one of the editors of the Bulldog Bulletin, the professional journal of the B-52 Gunner. This position also saw monthly TDYs to G units to maintain qualification in addition to helping roll out our new training programs.

In 1991 I was requested to join the 806th BMW(P) as they deployed to RAF Fairford for Operation Desert Storm as a staff planner and gunner. While there I helped plan, prepare and brief each days sorties. In addition I flew numerous combat missions as the lead gunner in the cell. When HQ SAC removed the gunners from the aircraft in October of 1992 I made my final B-52 flight with the 7th BMW that month and afterwards re-trained in the Flight Engineer field on C-141s. I attended flight engineer training at Altus AFB, OK. and was then assigned to the 30th ALS at McGuire, AFB, NJ. I took an early retirement at the end of 1993 to return to Alabama to operate the family business after the death of my parents in an aircraft accident.

I am an instrument rated pilot with over 1000 hours in the private sector, over 3,000 hours on B-52s and approximately 1,000 in C-141Bs. I am a franchise business consultant for Aaron's Sales and Lease, a job that takes me all over the country working with some great franchisees and associates. I have been residing in Lincolnton, NC. for the past fourteen years but dream of the ocean every day. I am happily married to my best friend, Dianne Greer, for over 17 years now and together we have five great kids (all grown), Brian, Michael, Aimee, Emory and Emilee. We also currently have three wonderful grandchildren, Elijah, Brayden and Noah. I hope one day that they will all understand the sacrifices made on their behalf by all of our service members, some who were not as fortunate as we to be around and able to tell our story.

Mike Jones

MSgt USAF (Ret.) Enlisted in the Air Force on March 12, 1957, and was assigned to B-52 Tech School at Chanute AFB, IL. His first duty assignment was at Walker AFB working on B-52Es. He became a Crew Chief on March 1, 1965, and was later assigned to Ramey AFB to work on B-52Gs. While there he was sent TDY to Guam for six months where he flew on 22 combat missions.

From 1970 to 1972 he was assigned to Beale AFB, CA, again with B-52G models. His next assignment was to U-Tapao as a Maintenance De-briefer, then Job Control. From 1973 to 1976 he was assigned to Rickenbacker AFB to work on KC-135s.

His final assignment was Andersen AFB, Guam, to work on B-52D models from 1976 until 1979 when he retired. He was awarded the Meritorious Service medal and three Air Force Commendation Medals.

Jay Lacklen

Lt Col, USAF (Ret.) is a retired Air Force Reserve Lt Col and pilot with over 12,000 military flying hours. He entered the service in 1970, retired in 2004, and flew in all conflicts from Vietnam to the 2003 Iraq War.

He flew 330 hours in the C-7 Caribou out of Cam Ranh Bay, RVN. He flew 2,000 hours in the B-52 from Loring AFB, ME, Castle AFB, CA, Andersen AFB, Guam, and U-Tapao AB, Thailand. He led the last Arc Light cell formation out of U-Tapao in June of 1975. Lacklen flew 9,500 hours in the C-5 Galaxy with the reserves from Dover AFB, DE and Westover AFB, MA where he performed as Operations Officer for Operation Desert Storm. After retiring, he taught pilot training simulators at Columbus AFB, MS.

Jay appeared twice on the CBS News program "60 Minutes" opposite the Air Force, in 1998 on TCAS procurement and in 2000 on the anthrax inoculation program. Lacklen is currently writing the second volume of a three-volume memoir, "*Flying the Line, an Air Force Pilot's Journey*," from his home in Vienna, VA.

Roland R. LaFrance, Sr.

Former SSgt, USAF, was born on July 22, 1954 at Nashua, New Hampshire. My father was a City firefighter. Graduated Bishop Guertin H.S., June, 1972. Inducted into the U.S. Air Force under delayed-enlistment program, April, 1972. USAF Basic Training, June to August, 1972, was honor graduate. Attended USAF Fire Protection Specialist School (57130) at Chanute AFB, IL. Graduating, October, '72.

Assigned to 42ND CES Fire Dept., Loring AFB, Maine. Rapidly completed training to become a 57150, Fire Protection Specialist. July, 1973: TDY assignment under "Bullet Shot" to Andersen AFB, Guam, housed in barracks area seven miles from base. Returned to Loring AFB after October Mid-East war shortened TDY.

1974: Met my future wife, Francesanne Loveless, daughter of an assistant fire chief. Promoted to E-4 (Sgt). Married in Feb., '75 at base chapel. As secondary rescue crew chief, became an Emergency Medical Technician-Ambulance.

Left active duty in June, 1976. Assigned to active Reserve firefighting flight, Westover AFB, Maine, promoted to SSgt/E-5. With hiring by the Nashua, Fire Department pending, I went to inactive status. Employed by the Nashua Fire Dept. six years.

Moved to Northern New York, 1984. Joined Saranac Vol. F.D., enlisted in New York Army National Guard as a medic (91A-20), rank of Sgt. Attended Combat Medical Specialist School in 1986, was an honor graduate. After two years went inactive, until Operation Desert Shield. Enlisted in Army Reserve, 310th Field Hospital, Malone, New York.

Active duty in 1991 - 92, attended college for LPN (91C-20) school, on President's list. 1993, as a lieutenant, and department chaplain, was injured battling house fire, reinjured working as a nurse. Due to injuries, left the reserves permanently, and was classified as permanently, totally disabled. Had numerous operations.

My wife and I live in Winter Haven, Florida. We have three grown children, one a 1Lt in the USAF, and four grandchildren.

Ted Lesher

Major, USAF (Ret.) Born and raised in an academic and flying family in Ann Arbor, MI, he earned his private pilot's license at age 16. Dropped out of University of Michigan Engineering School in junior year and through a convoluted set of circumstances entered Aviation Cadet Navigator Training at James Connally AFB, TX in 196. After Electronic Warfare Officer Training at Mather AFB, CA, he attended B-52 CCTS at Castle AFB CA, 1962-63.

His first operational assignment was 320th Bomb Wing, Mather AFB CA, 1964-67 - a very active period. Participated in early development and testing of conventional bombing equipment and tactics. He deployed to Guam when Arc Light first kicked off in February 1965, and flew Arc Light One in June. He, returned to Mather, and then completed second Arc Light tour in 1966.

Had three more Arc Light tours with 99th Bomb Wing, Westover AFB, MA, 1967-1970, then PCSed to 456 Bomb Wing, Beale AFB, CA. Beale initially marked a return to routine SAC alert crew duty, but in 1972 B-52 G-models started deploying to Southeast Asia under

Operation Bullet Shot and Beale became the Replacement Training Unit for this effort. Did much flying, training, and evaluating, and eventually volunteered for deployment and was present on Guam when Arc Light terminated in late 1973. He ended up with six deployments to SEA and 222 combat missions.

First 12 years in USAF encompassed the entire Vietnam era, when he rarely spent more than six months in any one place. Went PCS to Castle as an instructor in 1974 and as of 2016 hasn't moved since. This stability offered the opportunity to get married, resume education (BS, MS in Engineering Management from USC), get commercial pilot's license, and earn credentials in software engineering. After 15 great years and 7,500 hours in the B-52, he moved to a staff assignment with flight simulators and became deeply involved in software development at a time when simulators were first being computerized.

Retired from active duty in 1983, but came back in Civil Service in a SAC unit devoted to simulator support and supervised more officers and airmen as a civilian than as a military officer. Retired from that job when Castle closed in 1995 and went to work as a software engineer for a company that manufactures equipment for the communications industry. He retired for the third and hopefully last time in the year 2000.

He has taught many college courses as an adjunct at undergraduate and graduate levels. Owned two acrobatic airplanes and performed in airshows. He built and flies a 13-meter sailplane and took up skiing at age 50 and is still at it. He has traveled and hiked with wife and two sons to many exotic places around the world. In 2015 made a trip from California to Thailand without getting on an airplane, crossed 14 countries in five weeks including Russia and Mongolia on the Trans-Siberian Express. Would like to be retired but his wife, the wonderful Suki, an RN from Thailand, keeps him hopping.

Michael F. Loughran

Colonel, USAF (Ret.) Completed initial B-52 training at Castle AFB in 1969, then was a B-52H co-pilot in the 524th BMS, Wurtsmith AFB, for 13 months. A short tour to Vietnam flying the C-7 Caribou followed where he trained VNAF IPs. After Vietnam, it was back to Wurtsmith with a subsequent tour in 1CEG, then the first of three Pentagon assignments. He was the 69 BMS Ops Officer and Commander from Jan 83 to Mar 86. He attended Naval War College in

1987 prior to becoming the 2nd BMW DO and Vice Commander. Mike was the 416th BMW Commander at Griffiss AFB (Sep 90 - Jul 1993), during the Gulf war, the end of SAC Alert, the closure of SAC, ACC standup and the BRAC process which selected the 416 BMW for closure. He retired from the Air Force, worked in the defense industry and lives in Northern Virginia.

Frederick J. Miranda

Colonel, USAF (Ret.) grew up in Cohoes, New York. Graduated from St. Mary's Seminary & University, Baltimore, MD in 1962 with a Liberal Arts Degree.

Commissioned Second Lieutenant in May 1963 from Officer's Training School (OTS). Attended a 32 week Aircraft Maintenance Officer Course at Chanute. First assignment was to the 499th Air Refueling Wing, Westover AFB, MA. First TDY was a on a KC-135 Tanker Task Force escorting F-100s to Misawa; this turned into SAC Task Force Yankee Team supporting the first F-100 strikes out of Clark Field, Philippines into the Plain des Jars area just prior to the Golf of Tonkin incident. He learned more during that TDY than he did in 32 weeks of school.

Next assignment was a year at Tan Son Nhut, Vietnam, followed by a relatively short tour at 2AF Headquarters, Barksdale AFB, LA. Two years at 3Air Division which became 8th Air Force on Guam. This is where he became familiar with Arc Light and Young Tiger operations.

In June '70 he reported to SAC Headquarters, Maintenance Management Division, DCS/Logistics. Because of his Guam experience, he became the 'single point' for SAC Maintenance activities in SEASIA. Involved in Bullet Shot I thru V, Constant Guard and Linebacker I and II.

His most enjoyable assignment was to the 32d Tactical Fighter Squadron, Camp New Amsterdam, The Netherlands, where he served from '73-'77. There were about seven former guests of the Hanoi Hilton in the organization and they related, first hand, the joy and elation felt when the B-52s struck Hanoi for the first time in December 72.

Naval Command and Staff Course was next and then an opportunity to command the 319th Field Maintenance Squadron at

Grand Forks AFB, ND. He was pulled back to Offutt to work on the SAC IG Team from '80 – '82. He moved from the IG to the LG and was involved in planning for the B-1B. Last assignment was as Deputy Program Manager for Logistics in the B-1B System Program Office at Wright Patterson. Fred retired in May 88.

His decorations include: Legion of Merit, Bronze Star, Meritorious Service Medal with 3 oak leaf clusters, Air Force Commendation Medal with 1 oak leaf cluster, the Republic of Vietnam Campaign Medal with 3 campaign stars and the South Vietnam Cross of Gallantry with Palm.

Enjoyed a second career selling repairs for an FAA Approved Repair Station to major airlines at home and abroad. Presently attempting to make something of a terrible golf game – it's hopeless.

Joe Mathis

Former Captain, USAF, 1990-1999. Graduated from OTS at Lackland AFB in Aug. 1990, then attended Undergraduate Navigation School at Mather AFB, CA and B-52 training at Castle AFB, CA. Navigator in B-52H at K.I. Sawyer AFB MI, 644th Bomb Squadron, 1992-1994. Then served as Navigator and later as Radar Navigator in the 20th Bomb Squadron, Barksdale AFB, LA. Now works in the banking industry and lives north of Atlanta, GA.

Paul Paris

Major, USAF (Ret.) retired Air Force B-52 pilot with over 3,500 hours in the BUFF. I hold a B.A. in Geography from the University of North Texas and an M.A. in Management from Webster University. My second career was in the banking industry and I retired in 2007 to write books.

I won a United States Air Force writing contest with an article that appeared in *Road and Rec*, an international military magazine in 1990. My non-fiction short story was selected for inclusion into A Treasure Box, a book published by the North East Texas Writers Organization. My first book, *It's All Good, Life is an Adventure*, was my autobiography written for my children. *Amazing Faith, A Walk with Cancer* details how faith carried my family through the loss of my wife after a four-year battle with breast cancer.

In 2014, I completed *The Parachute Queen of Texas*, the biography of Almeda Babcock, a famous barnstormer in the 1930's. This year I finished my first historical novel, *Flight Risk, and The Flying Circus*, a non-fiction book for mid-grade readers. This month I finished co-writing my first children's book, *Brodie the Biplane*, and am in the process of submitting it to literary agents

George Schryer

SMSgt, USAF (Ret.) grew up in a small town in Ohio and after graduating high school in 1958, joined the US Navy. He spent five years in the Navy, three of which were aboard the destroyer USS Mullinix DD 944 as a Gunners Mate. In 1962 he switched to the Air Force and in 1966 became a Tail Gunner on a B-52 Bomber. He served two tours in South East Asia flying combat missions over South and North Vietnam from Guam, Okinawa and Thailand. In 1977 he was transferred to the US Air Force Survival School in Spokane Washington as a Staff Instructor, and retired in 1980 at the rank of Senior Master Sergeant.

His decorations include the Distinguished Flying Cross, Air Medal with two Oak Leaf Clusters, Purple Heart, Meritorious Service Medal, Vietnam Campaign Medal with three Stars and the Vietnam Service Ribbon among others.

He is a life member of the Veterans of Foreign Wars, currently holding the position of Commander of the Washington VFW Post 6088 and Commander of VFW District 2 for the State of NC, life member of the Disabled American Veterans, and the Air Force Gunners Assoc. He is also a member of the Vietnam Veterans Assoc. and the Jr. Past Governor of the Washington Moose Lodge and a member of the Beaufort County Committee for Constitutional Studies.

George has been married to Gail E. (Simmons) Schryer for 40 years, and has two daughters, five grandchildren and one great grandchild. He is retired and has lived in "Little" Washington NC since 1994.

Tommy Towery

Major, USAF (Ret.) earned a B.S. degree in Journalism in 1968 from Memphis State University where he also earned his commission as a 2nd Lt. through the AFROTC program. Attended Navigator

Training and Electronic Warfare Officer Training at Mather AFB, CA. Following B-52 CCTS training in B-52F models at Castle AFB, CA he was assigned to the 20th Bomb Squadron at Carswell AFB, TX flying B-52C and B-52D models. He was deployed for six months to Guam as part of "Operation Bullet Shot" and was assigned to 8th Air Force Bomber Operations as an Arc Light mission planner. On his second deployment he worked as an 43rd Bomb Wing Arc Light mission planner and flew on B-52 combat missions as a staff officer and helped plan Linebacker II missions. In 1976 his crew received the Mathis Trophy, awarded to the top bomber unit based on combined results in bombing and navigation in the SAC Bomb-Nav Competition.

He logged over 1,600 hours in B-52s and over 5,000 hours total Air Force flight time. His decorations include the Meritorious Service Medal with oak leaf cluster, the Air Medal with eight oak leaf clusters, the Armed Forces Expeditionary Service Medal with cluster, the Vietnam Service Medal, the Republic of Vietnam Campaign Medal and the South Vietnam Cross of Gallantry with Palm. He retired from the University of Memphis in 2008 and lives in Memphis, Tennessee, with his wife Sue who graciously allows him to devote many hours to his writing hobby.

Tommy has written five non-military books, *"A Million Tomorrows –Memories of the Class of '64," "While Our Hearts Were Young,""The Baby Boomer's Guide to Growing Up in 'The Rocket City'," "When Our Hearts Were Young – Today, 50 Years Ago," and "Goodbye to Bob."* He is a writer and editor of six other books in the "*We Were Crewdogs"* series and a contributing editor and publisher of a revision of *"Linebacker II – A View from the Rock."* He also published *"Moments of Stark Terror!"*dealing with B-52 emergency events and *"We Were Crewdogs – The Vietnam Era"* which is out of print. He is a member of The American Author's Association and The Military Writers Society of America.

Steve Winkle

Major, USAF (Ret.) is a retired B-52 Instructor Radar Navigator. Prior to attending Undergraduate Navigator Training (UNT), he was a Minuteman II Deputy Missile Combat Crew Commander and Missile Combat Crew Commander. Following UNT, he attended Navigator-Bombardier Training. He was assigned to B-52Ds at Carswell AFB, TX and went TDY to Southeast Asia where he flew 62 combat missions. Following his retirement from active duty, Steve got involved in telecommunications training and retired from the corporate world as Vice President of National Training for Allegiance Telecom, Inc. He also served one term as a member of the Bedford, TX city council where he currently resides. Linda, his wife of 46 1/2 years recently passed away.

Randall E Wooten

Colonel, USAF (Ret.) Randy is a1968 ROTC graduate from East Texas State University, now Texas A&M, Commerce. He attended UPT at Randolph AFB, Texas, and graduated in class 70-03 in October 1969. We was assigned to Viet Nam as a Forward Air Controller in the O-1E Bird Dog, and served in Duc Hoa and Tay Ninh. In December 1970, he received his B-52 assignment to Dyess AFB, TX, as a co-pilot, and was in the 337 BMS, 96 BMW, and was assigned immediately upon check-out to Stan Eval. He deployed in February 1972 for operation Bullet Shot, and Linebacker.

During his 29-year career, Randy had additional B-52 assignments to Andersen AFB, Guam, Barksdale AFB, Louisiana, and Blytheville (Eaker) AFB, Arkansas. He also served at SAC HQ from 1982-85, working in XPPB (plans). Additionally, he had three tours at Air University—SOS faculty, ACSC, AWC, and finished his career at AU as Commander, Ira C. Eaker College for Professional Development; Commander (and founder) College for Enlisted PME; and Director of Plans and Operations.

He has 309 combat missions as a FAC, and in B-52s in SEA and Desert Storm (Fairford, England). Randy has over 5,000 hours of flying time, and currently flies his Cozy MK IV, experimental airplane for fun. He and his wife Gayle live in the Houston area, and have two children, Amber and Alan; two step-children, Heather and Mitchell; and two grandchildren, Levi and Morgan.

Jim Wuensch

Major, USAF (Ret.) entered the Air Force in 1982 and was commissioned via OTS. After UNT at Mather and CCTS at Castle, he was assigned to the 20th BS at Carswell. In 1986 he took a position as a UNT instructor in the 451st FTS at Mather. Jim then PCS'd to KI Sawyer where he worked in Stan/Eval and Bomb/Nav. After K.I. Sawyer Jim escaped south to be the AFROTC Commandant of Cadets at the University of Texas. From there it was off to a Joint position near Washington DC working in the unacknowledged black world.

After his first miss to Lt Col, the AF sent him back to fly at Minot, where he was sure it was going to be his last three years in the AF. In May of 2001 the AF invited him to stay an extra four years with a $15,000 annual bonus which he couldn't resist. He finally retired in 2006 and took a console operator position in the B-52 simulator at Minot in 2007 and worked there until last year.

John York

Lt Col, USAF (Ret.) was a distinguished graduate of Officer Training School and attended Undergraduate Pilot Training at Laredo AFB, Texas. He survived a year in Vietnam flying the highly subsonic O-1E. He flew the B-52G at Barksdale AFB, LA, Anderson AFB, Guam, Barksdale AFB, LA, Wurtsmith AFB, MI and Barksdale AFB, LA. He also flew the T-39 and KC-10. In retirement he flew at FedEx Express (Scumbag Freight Dog) until he got so old the FAA wouldn't let him fly anymore. John now operates a Jackass ranch in rural Oklahoma. His wife of 48 years, Ruthie, says "He is with his own kind now."

Personal Notes

We Were Crewdogs – The Series

Tommy Towery

We Were Crewdogs I – The B-52 Collection

We Were Crewdogs II – More B-52 Crewdog Tales

We Were Crewdogs III – Peace Was Our Profession

We Were Crewdogs IV – We Had to Be Tough

We Were Crewdog V – We Flew the Heavies

We Were Crewdogs VI – Freedom Is Not Free

Many stories have been written about the B-52 combat missions flown in Vietnam. There's also many books on the B-52 aircraft. On the other hand, there have been very few things written about the day-to-day training, peacetime missions, and everyday life of a B-52 Crewdog during the Cold War era or recent wars.

When the Strategic Air Command ended its Nuclear Alert readiness in 1991, many of us were already out of the B-52 and retired from the service. Now it seems even longer in the past, and not only do aircrews on Nuclear Alert no longer exist, neither does the Strategic Air Command, the command in which many of us spent the majority of our military service.

That's the educational goal of this collection of stories. Some people sit down and write stories to educate others. Some people write stories to entertain. Some record history. We hope that in our endeavors, this group of stories will do all of these things. It is a collection of the serious and the humorous sides of living the life of a SAC B-52 Crewdog, and it is intended to share those experiences with others.

The common thread that binds them is the B-52 aircraft - the BUFF (Big Ugly Fat Fellow).

Moments of Stark Terror!

Tommy Towery

Flying in a B-52 is hours and hours of pure boredom, followed by moments of stark terror!

If you are not interested in the peacetime alert or the non-flying tales of B-52 crewmembers and are only interested in the edge-of-the-seat, fingernail-biting stories of men under extreme pressure, then this book is the one you want to read.

It includes the complete story of the Red Cell mid-air out of Guam. It also includes the shootdown of a B-52 over New Mexico by a F-100 fighter. There are multiple stories of ejections and survival, both in combat and peacetime environments. One story is told by a crewmember who ejected from a B-52 and survived, even after riding his ejection seat to the ground without a parachute. There are stories from the first crew to be shot down over Hanoi and the last B-52 shot down in the Vietnam War.

All the stories are about the mighty BUFF, the B-52 Stratofortress and the men who flew it, and their personal moments of stark terror.

*** D I S C L A I M E R ***

All the stories in this book have previously been published in Volumes 1-6 of the "We Were Crewdogs" series. This book is a collection of only the stories that involved inflight emergencies, ejections, crashes, and combat moments involving a sense of danger and the rapid responses needed to control the situation.